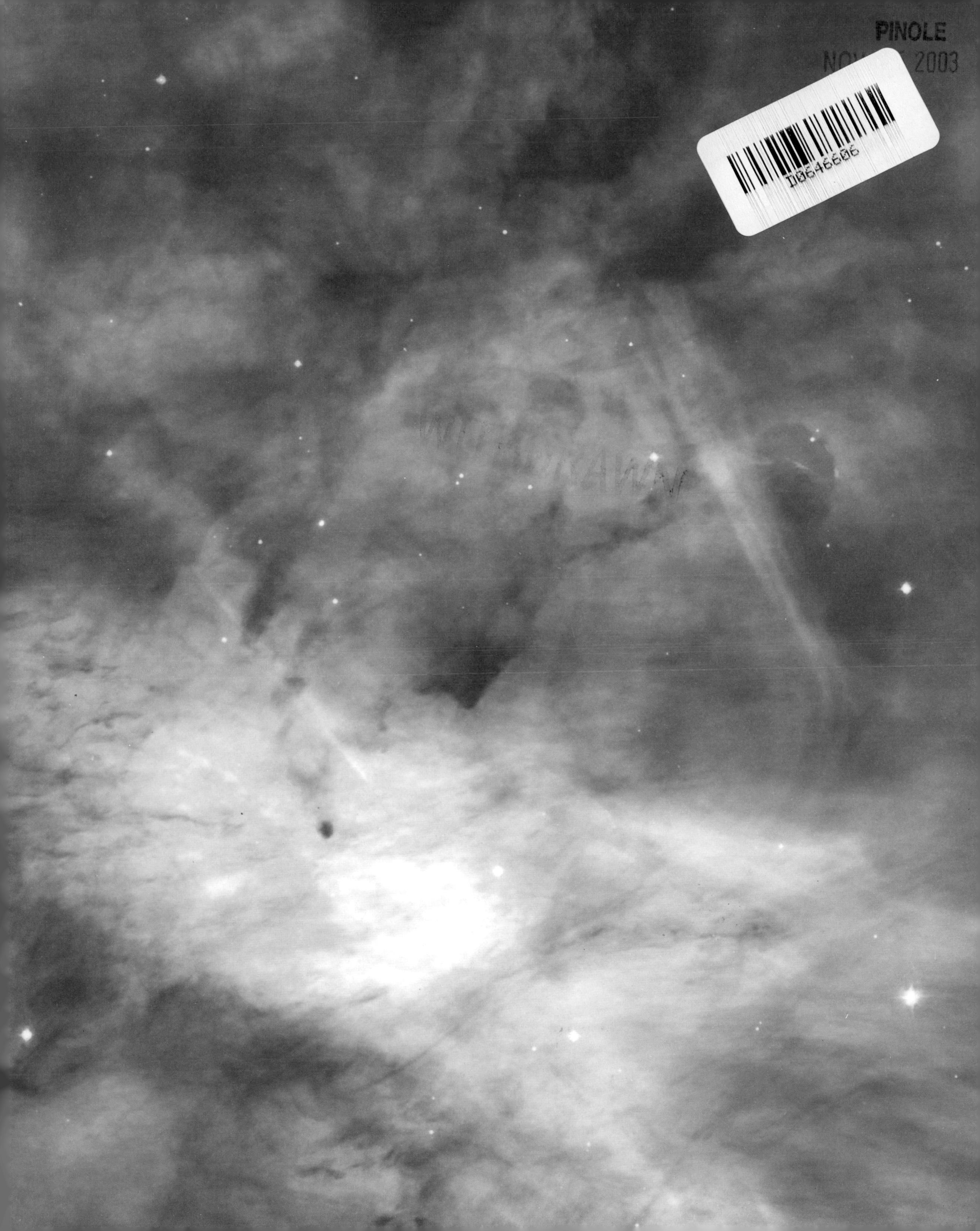

UNDERSTANDING
the UNIVERSE

With loving dedication to Kamini and Teresa,
to our parents, and to our precious children.

We are pleased to thank Caroline Rayner and Frances Adlington
at Philip's for their efforts in creating this wonderful book.

Understanding the Universe
Copyright © 2002 Raman K. Prinja & Richard Ignace
Copyright © 2002 by Philip's, a division of Octopus
Publishing Group Limited

Checkmark Books
An imprint of Facts On File, Inc.
132 West 31ˢᵗ Street
New York NY 10001

For Library of Congress Cataloging-in-Publication data,
please contact Checkmark Books. ISBN 0-8160-5228-X

Checkmark Books are available at special discounts when
purchased in bulk quantities for businesses, associations,
institutions, or sales promotions. Please call our Special
Sales Department in New York at (212) 967-8800 or (800)
322-8755.

You can find Facts on File on the World Wide Web at
http://www.factsonfile.com

MANAGING EDITOR	*Caroline Rayner*
COMMISSIONING EDITOR	*Frances Adlington*
EXECUTIVE ART EDITOR	*Mike Brown*
DESIGNER	*Alison Todd*
PICTURE RESEARCHER	*Cathy Lowne*
PRODUCTION CONTROLLER	*Man Fai Lau*

Printed in China

10 9 8 7 6 5 4 3 2 1

This book is printed on acid-free paper.

UNDERSTANDING
the UNIVERSE

RAMAN K. PRINJA & RICHARD IGNACE

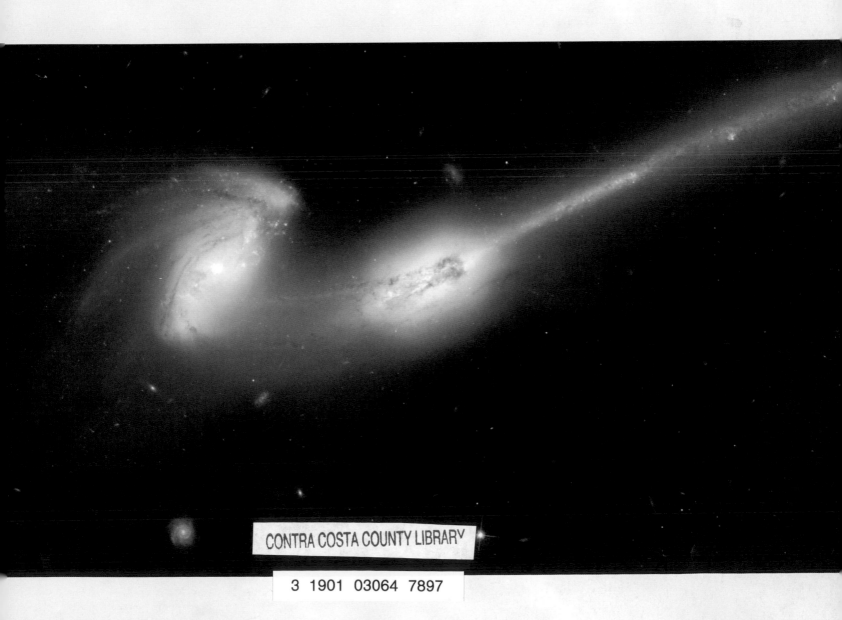

Foreword

Our knowledge of the Universe is increasing with leaps and bounds with every passing decade. It is hard to realize that less than a hundred years ago astronomers were still arguing about the nature of the gaseous nebulae, with many believing that the spiral nebulae were planetary systems in the making. What would our ancestors of those days make of the "Hubble Deep Field," a keyhole view of the Universe, photographed with a telescope orbiting the Earth, showing galaxies seen at a time when the Universe was only about 10 billion years old? What would they make of gravitational lenses or the close-up images of the surface of Mars from the Mars Orbiter spacecraft?

But there are many fundamental questions still to be answered and one of these has occupied human curiosity since time began. Are we alone in the Universe? Is planet Earth, orbiting our star, the Sun, the only place in the Universe to support intelligent life? In recent years many planetary systems have been discovered around other stars; we watch with excitement to see where this research will lead.

Understanding the Universe brings these stories right up to date and is beautifully illustrated with some stunning images. It is written for the general reader by professional astronomers closely involved in current astronomical research. We are taken on a voyage of discovery through the cosmos with the book as our travel guide. Our essential travel documents are included, giving us access to places we may visit by helping us to understand the background to this fascinating subject. As we pass along we find out about the current fields of research which are intriguing today's astronomers.

If you are interested in finding out what we currently understand about the Universe I am sure you will enjoy this book.

Dr. Margaret Penston, MBE
President, SOCIETY FOR POPULAR ASTRONOMY

Contents

Welcome to the cosmos

1

Some of the most remarkable and breathtaking discoveries of the Universe are presented in this book, revealing our understanding of the planets, stars and galaxies. Astronomy today is an exciting and fascinating subject to a broad spectrum of people. The study of the Universe uniquely combines curiosity, discovery and exploration, and thus appeals to the basic instincts of human nature. This is an exceptionally exciting time for the exploration of space. Technological advances have provided us not only with probes that can visit other planets, but also with giant powerful telescopes that may be located on high mountain tops or even in orbit around the Earth. These excellent facilities provide us with a whole new perspective on the beauty and workings of the cosmos. The aim of this book is to provide a first class tour of this marvelous Universe.

Your passport to the Universe

In this opening chapter, we introduce the subject of astronomy, and the basic terminology and concepts that are repeatedly used. It is especially important to grasp an understanding of the huge scale of the Universe and its constituents. We often have to deal with a vast range of distances and sizes not normally encountered outside the realm of astronomy. These first steps represent the "passport" that will allow us to embark on and enjoy a journey through the planets, stars, galaxies and beyond. So, sit back, relax and take in the sights! The itinerary consists of trips to majestic places throughout the Universe, learning about some remarkable discoveries, and viewing some outstanding images.

A brief tour of the Universe

The first step to making travel plans is to view the brochures. What are the main attractions? What are the travel arrangements? On this trek through the cosmos, liftoff is from planet Earth, a seemingly insignificant mote (except for the fact that it is our dwelling place!) circling a rather average star in a non-descript galaxy that is literally just one of billions. Fasten your seatbelts as the journey begins, launching you upward through the cloud-tops and away from the Earth's atmosphere into the frigid and near vacuum conditions of space.

The first leg of the tour is a grand procession through the Solar System, including the best views of the resident planets. The Earth's siblings are varied indeed, ranging from massive Jupiter, with its crushing gravity and high-speed winds, to Venus, which, although a near twin of Earth, is entirely inhospitable for life due to its torrid and acidic atmosphere, and to tiny Pluto, which is an iceball at the edge of the Solar System. There will also be detours to visit spectacular comets and pitted asteroids.

The next leg of the journey will bring us into the mesmerizing fields of stars. These rather innocuous pinpricks of light as seen with the naked eye from afar are powerfully and breathtakingly majestic when viewed through large telescopes. The nearest star, our own Sun, is known intimately, and stars throughout the Universe vary greatly in size from a hundred times smaller than the Sun to a hundred times greater. In terms of radiant power, stars range from one millionth of the Sun's output to over a million times more. Many stars do not exist in isolation but come in pairs (binaries) or multiples that are locked together by their mutual gravity, and in some cases stellar atmospheric material is drawn from one star to the other. As well as all this, stars can have equatorial disks, high-speed jets, and ferocious outflowing winds. Finally on this part of the journey, and most spectacularly, violent explosive behavior among the stars will be seen, the remnants of which may be strange and unusual stellar embers or even black holes.

In the final leg of the trip, the grandeur of the cosmos on the large scale is revealed. Giant, swirling spiral galaxies seen in exquisite detail are among the jewels of the observable Universe. However, not all galaxies are so peaceful. In fact, many exhibit extreme violence, with collisions, mergers and galactic cannibalism. In deepest space, at the farthest reaches of the observable Universe, powerful galaxies called quasars are known to exist, harboring supermassive black holes and spewing material out to form high speed jets. Permeating the entire Universe, from the Earth to beyond the quasars, is the cosmic background radiation, which is a low-level radiation bath existing at microwave frequencies. This radiation is a remnant from an explosion of cosmic proportions at the beginning of time.

So, you have read the "brochure" and now you are ready to go. All that remains is the "passport," which in our context is some essential background information that will make this

◀ *The most familiar star is the Sun, which completely dominates our Solar System. It is far too hot to have a solid surface, and all the atoms and molecules contained within it are in the form of gas. The Sun is shown here in special light to highlight the bright (hotter) and dark (cooler) patches of gas. A gaseous prominence can be seen at upper right.*

journey successful. We begin with a discussion of the nature of astronomical inquiry and then progress to a description of observing objects in the night sky.

What is astronomy?

Astronomy is an incredible field of study for many reasons, not least of which is its accessibility to anyone who wanders out on a clear night and gazes skyward to the stars. The sky is the laboratory, observation is the experiment, and eyes are the equipment. A significant amount of basic astronomical science can be accomplished without any observing device other than the human eye. That is how the Arabians, Chinese, Greeks, Indians and Native Americans did it in ancient times. The chief subjects of this book, however, are those things in the Universe that are not discernible with the naked eye.

More work has been done and a greater scientific understanding gained of the Universe and its contents in the last century, and arguably in the last three decades, than in all of the time preceding. The advent of large ground-based telescopes, such as the giant Keck telescopes at Mauna Kea in Hawaii, and especially of orbiting satellites, such as the Hubble Space Telescope, have ushered in an unprecedented era of active and fruitful study of virtually every possible topic in astronomy. Before delving into the

many interesting, and even surprising, results of recent years, we must first introduce the subject of astronomy together with some of the concepts and jargon that will be common throughout most of the discussions.

A good starting point is to ask the basic question, "What is astronomy?" The answer is not particularly straightforward. Studies relating to the physical properties of the Earth, its interior and atmosphere, broadly fall into the realms of geology and geophysics. One could therefore think of astronomy as the study of everything not of the Earth. Alternatively, the word "astronomy" literally means the apparent positions of things in the heavens. Although positional astronomy dominated the subject up until the 20th century, modern astronomy encompasses far more than the motions of heavenly objects in the sky. It would actually be more correct to use the word "astrology" to describe the study of the heavens, since the suffix "-ology" refers to "the science of." However, the reader should be well aware that astrology has taken on a specific cultural meaning, relating to how the stars, planets and Moon allegedly influence human affairs; it is not at all related to the science of astronomical study.

Today, most would agree that astronomy is really a subdiscipline of physics, meaning that astronomy is a physical science that concerns the measurement and interpretation of data. Astrophysics is, therefore, the most accurate descriptive term of what astronomers do, and it is the term used by many to describe their profession. In the end, perhaps the best answer to "What is astronomy?" is simply that it is the physics of the Universe. But astronomy differs from other scientific pursuits, such as geology, biology or chemistry, in a significant way. When, for example, experimentalists decide to study the effects of radiation on, say, plant life, they gain the use of a laboratory, assemble some equipment, perform a variety of planned experiments, record and study the results, and draw conclusions about the subject and possibly some predictions for future study. But this is not so with astronomy. Astronomers have not designed

▼ The Keck telescopes are among the largest in the world, with each having an aperture spanning 10 m (394 inches). This picture shows the observatory domes that house the giant telescopes on the extinct volcano of Mauna Kea in Hawaii.

the experiment. They did not start it; they do not know how to stop it; and they certainly have no control over it. In fact, humans and the Earth are part of the experiment, indeed a very small part in most respects. Except for objects in our own Solar System, such as the Moon or the occasional passing comet, astronomers cannot take samples from their objects of study or even get a view of the Universe from a different location.

Astronomy is, therefore, an observational science in which perspective is a key consideration. Except for the Solar System, astronomers are constrained to study only the light from distant objects. Referring back to the laboratory analogy, the goal of the astronomer is to infer the nature of the experiments that are taking place, how long they have been running, and what the settings are for all the "knobs." Unlike terrestrial scientists, astronomers cannot ask "What happens if we change this?" but rather "If we see this, what has happened?". It is a very different situation indeed.

Even so, it is amazing how far astronomy has come, as will be evidenced by this book. There are three main factors that contribute to this success. First, there are a lot of different objects in the Universe. For example, because stars change comparatively little except over lengthy periods of time, it would be difficult to understand stars if there was only one to study. But in fact there are very many stars to study, and they are all of different ages, sizes, masses, temperatures and luminosities (a term used by astronomers to describe the amount of power output, especially in the form of light). By looking at stars as a group, astronomers have been able, after a century of hard work, to piece together a fairly coherent picture of how stars form, live and die.

Second, the physics governing the principal behavior of many astronomical objects can be somewhat simple: the Universe is largely gaseous; stars are basically spherical; the motions of stars

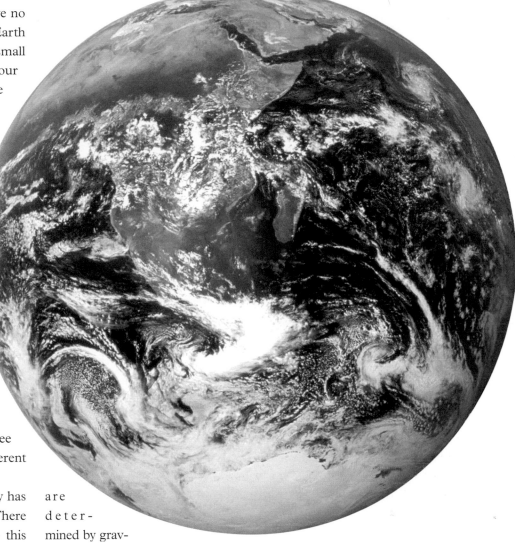

are determined by gravity; and so on. Of course, the details are very interesting (and sometimes the details turn out to be of extreme importance), but a simplistic approach helps astronomers to understand or interpret the basic properties and behavior of the great variety of objects that exist throughout the Universe.

Third, there have been numerous significant advances in technology, especially over the last half of the 20th century. These advances range from the speed of computers, to techniques for building larger and better telescopes (especially for orbiting satellites), to the development of devices for measuring and recording light from stars and galaxies. Every significant advance in astronomical observing technology has led to interesting and important new discoveries about the Universe, and since technological break-

▲ *This photo of the Earth was taken by the Clementine spacecraft in 1994 from orbit around the Moon. This "photo" is actually a composite of 70 separate images.*

throughs just seem to keep on coming, continued surprises in astronomy can be expected for the future.

Why study astronomy?

But why should anyone be interested in astronomy? What is its attraction? The answers to these questions vary between individuals, but here are a few possibilities. First, there is the basic drive of curiosity that people have always had: how does it work and why? It is like climbing a mountain: one does it because it is there; it represents a challenge and a thrill; it is something to be conquered; it is an achievement.

Second, astronomy involves some of the "big questions." Is there life elsewhere in the Universe? Does the Universe have a beginning or an end in time? Is it truly infinite in size? Are there other Universes?

Third, certain phenomena in astronomy are "extreme" in the physical sense and cannot be reproduced on the Earth. Outer space is by far a better vacuum than any that has been produced on the Earth. Black holes are believed to exist in the centers of many galaxies and in some stellar binaries (a binary consists of two mutually orbiting stars). These black holes are points of tremendous gravitational influence, so strong that not even light can escape from them. Nuclear fusion occurs in the hearts of stars, transforming hydrogen and helium into heavier elements like carbon, oxygen and iron. Some stars undergo supernova explosion, one of the brightest and most energetic events in the Universe.

A sense of scale

One of the truly daunting aspects of astronomy is the incredible range of scale. This term refers to how large or small a number may be to describe some property. For example, a young child is around 1 m (about one yard) in height, whereas a tall building might exceed 100 m. Here the heights of the child and building are being contrasted. The unit of description is the meter, and the scale for children is "small" relative to the building which is "large."

Astronomers describe objects and events in terms of numbers and units. The most basic information of astronomical interest consists of how far (distance), how large (size), how much (mass, energy), how old (time), how fast (speed), how hot (temperature), and how luminous (power). Astronomers are certainly interested in other kinds of information, such as magnetic fields and rotation speed, but the preceding list represents some of the most important descriptive properties of planets, stars and galaxies.

Since the study of astronomy in its simplest sense involves everything not of the Earth, the

Scaling it down

One way to bring astronomical scales down to a more familiar level is to consider analogies. Here are a few simple examples:

The Sun

The Sun is a rather typical star. Thinking of the Sun as being the size of a football...

...the Earth would be pea-sized or smaller

...Jupiter would be the size of an infant's fist

...and the largest stars would rival a sports stadium

Distances between the stars

The distances between stars are truly immense, and a scale model can be valuable for understanding their span. Imagine that the distance from the Earth to the Sun was roughly the length across a person's palm...

...then the distance to the nearest star would be about the breadth of the English Channel at its narrowest point

...and the distance to the center of the Galaxy would be equivalent to more than four circuits of the Earth's equator

The Milky Way and distances between galaxies

Most of the stars (including our Sun) and gas in our Galaxy, the Milky Way, reside in a flattened disk or pancake-like region. Suppose the thickness of the Milky Way was equal to the thickness of a typical sheet of paper...

...the distance from the Milky Way to one of the nearest galaxies, the Large Magellanic Cloud, would be like a stack of 170 sheets of paper

...the distance to the nearest big galaxy like our own, the Andromeda, would be a stack of about 2200 sheets, or over 4 reams

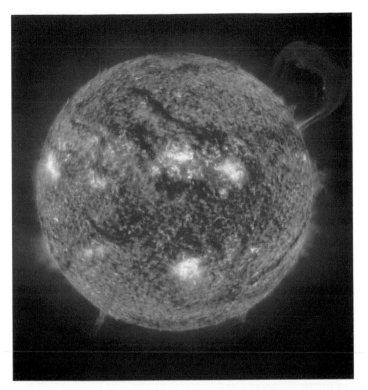

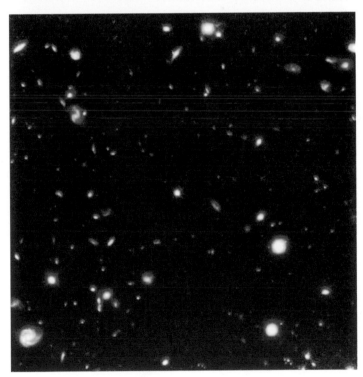

▲ Objects in the Universe span a diverse range of sizes, from tiny comets to colossal galaxies. Displayed here is a selection of such objects at different scales. The Sun again (top left), now showing the extended glowing coronal region that envelopes the Sun and a prominence at upper right. Prominences can rise out to thousands of kilometers, ejecting material that traces out the looping magnetic fields of the Sun. Next is the red supergiant star Betelgeuse (top right), which is about 1000 times larger than the Sun. Betelgeuse is the bright reddish star that marks the upper left shoulder in the constellation of Orion. Following is a Hubble Space Telescope image of the grand design spiral galaxy NGC4414 (bottom left) at a distance of 60 million light-years. Last is an infrared view (bottom right), showing some of the faintest and farthest galaxies ever observed, some of which are as distant as 12 billion light-years.

sheer range in numbers used to describe astronomical occurrences is truly stunning. The numbers range from sizes and masses that are atomic to tremendously greater values that describe the Universe as a whole. Grams, centimeters and inches, and kilometers and miles are measurements that are familiar to the reader, but what about light-years and solar masses? Sometimes in astronomy the numbers are so large that even the powers of scientific notation can become bulky if units common to terrestrial use are adopted. For example, the Sun is about 150 million km (or 1.5×10^8 km), which is 93 million miles (or 9.3×10^7 miles), from the Earth. This is a large number certainly, but the next nearest star is about 300,000 times farther away, or 4×10^{13} km away (about 2.5×10^{13} miles). And stars near the center of our Milky Way galaxy are yet another 7500 times farther away still, or about 3×10^{17} km (2×10^{17} miles).

Astronomical distances tend to be so large that kilometers and miles are not especially useful for measuring them. Astronomers have therefore defined other units of measure for the sake of convenience and to make the meaning of the large numbers easier to grasp and appreciate. Two commonly used measures for large distances are the astronomical unit (AU) and the light-year (ly). One astronomical unit is the distance between the Earth and Sun. The AU is therefore appropriate for discussing interplanetary distances. For example, the closest planet, Mercury, is only 0.4 AU from the Sun, whereas the farthest planet, Pluto, is about 40 AU from the Sun. For interstellar distances, the light-year is used, with 1 ly being the distance traversed by light, traveling at the speed of light, over a period of one year. Since light travels at a speed of about 300,000 km/s (186,000 miles/s) and since there are roughly 30 million seconds in a year, 1 ly is about 10,000 billion km, or 1×10^{13} km (6×10^{12} miles). So the nearest star is about 4 ly away.

However, even in light-years, the nearest galaxies, the Large and Small Magellanic Clouds (LMC and SMC), are at distances of about 170,000 ly, and the nearest big galaxy (the Andromeda Galaxy) is over 2 million ly away.

Big and small numbers

Numbers like the mass of the hydrogen atom at 0.00000000000000000000000000167 kilograms, or that of the Earth at 5970000000000000000000000 kilograms are awkward in their being extremely tiny or huge. A shorthand approach to writing such numbers is called scientific notation. The idea is that all the zeros act as "padding" to get the relevant digits in the correct powers-of-ten location. For example, consider the decimal number 0.002084. In scientific notation this becomes 2.084×10^{-3}, the superscript number (or "exponent") indicating how many places the decimal point must be moved to recover the longhand representation, with *negative* powers indicating small numbers (zero padding to the left) and *positive* powers indicating big numbers (zero padding to the right). So the hydrogen mass becomes 1.67×10^{-27} kilograms, the -27 representing 26 zeros to the left of 167 without having to write them all out. The Earth's mass becomes 5.97×10^{24} kilograms. Now the 24 refers to the number of places between the first digit, 5, and where the decimal point would appear. In this case that would mean 22 zeros following the 7. The convenience of scientific notation means that we no longer need to write out long strings of zeros, and that by simply looking at the exponent – whether it is positive or negative, big or small – we can appreciate the scale of the value of interest.

When numbers become this large, a further convenience is to use the various metric prefixes, such as mega (for 1 million or 1×10^6) light-years or Mly for distances between galaxies, and even giga (for 1 billion or 1×10^9) light-years or Gly for cosmological distances, referring to the scale of the whole Universe.

It is worth noting that the light-year gives immediate relevant information about an object: if a galaxy is at a distance of 1 Mly, the light being observed today originated from that galaxy 1 million years ago. The fact that light travels at a finite speed means that astronomers view distant objects as they were at the time the light was emitted, hence astronomical observation allows us to see the Universe as it was in the past.

Viewing the night sky
The Constellations - a traveler's guide
From ancient times the constellations have represented familiar patterns of stars in the sky,

► *The constellation of Orion is prominent in winter skies. Betelgeuse is the reddish bright star at upper left and Rigel the bluish white bright star at lower right. The three relatively bright stars falling in a straight line mark the Hunter's belt, making this constellation fairly easy to identify. An arc of several bright stars to the right of the Hunter serves as his bow, which he is aiming toward Taurus, the Bull.*

Location, location, location

Constellations are useful, but a system for exactly pin-pointing the positions of objects in the sky is necessary, as for example when pointing a telescope. The coordinate system that has been chosen by astronomers is quite similar to the system of longitude and latitude used for the Earth, except that for sky coordinates (or the "celestial sphere") astronomers use the terms Right Ascension (RA) and Declination (Dec.).

Earth longitude is measured in degrees east or west from the prime meridian, which passes through Greenwich, London, England. RA, however, is measured in units of angular time (hours, minutes and seconds) rather than in angular degrees (degrees, arc minutes and arc seconds). RA is measured eastward from a special position on the sky called the point of Aries. The point of Aries marks the crossing of the Sun and the celestial equator (the projection of the Earth's equator on to the sky) that occurs on the Vernal Equinox (the first day of spring).

Declination is very similar to Earth latitude, being measured in degrees from the equator. The only difference here is that latitude is measured in degrees north or south from the Earth's equator, whereas Dec. is measured in positive or negative degrees from the celestial equator. For example, the Earth's north pole is at 90 degrees north (90°N) latitude, whereas the north celestial pole is at +90 degrees (+90°) for Dec.

An example of RA and Dec. is the point of Aries itself, which is zero for both and is written as RA = 0^h 0^m 0^s and Dec. = 0° 0′ 0″. Since there are 24 hours in a day and 360 degrees in a circle, 1 hour of time measured in RA is 15 degrees of angular measure. But why is RA measured in time? Astronomers often want to know when an object, such as a star, will be most nearly overhead. Since stars rise and set owing to the Earth's rotation, it is convenient to measure RA in time in order to determine more easily the rising and setting time of an object, and how long before it passes overhead.

leading mariners and explorers through unknown places around the globe. Constellations are probably the things most readily associated with astronomy, and almost everyone knows the names of a few constellations. These star patterns, most of which represent animals or heroes from mythology, remain a useful means of dividing up the night sky into conveniently sized patches or areas.

The Zodiac is perhaps the most well-known set of constellations, of which there are traditionally 12 in total dating back to before 2000 BC. (Actually, there are 13 major constellations constituting the Zodiac.) A standard set of 88 constellations was adopted in 1922. Of these, more than half had first appeared in the *Almagest*, a book published in approximately AD 150 by the Greek astronomer Ptolemy (who was better known for his attempts to describe the motions of the planets using constructions based on circles). Ptolemy's list was amended over the years as European explorers began traveling to regions where stars near the South Pole could be observed.

For both the professional and the amateur astronomer, constellations are useful for directing attention to different parts of the sky.

▶ *This majestic photograph taken on Maui (Hawaii) on 3 March 1999 shows a procession of four prominent planets near sunset: at top is Saturn, then bright Venus, and then Jupiter, with Mercury nearest to the horizon.*

Constellations also form a basis for naming objects, such as the brightest stars, that fall within a particular area. There are many brilliant stars scattered over the entire night sky, but within a given constellation, there will normally be only a few stars that substantially outshine their neighbors. Although all of the very brightest stars have proper names given to them by the Greeks, Romans or Arabians (for example Sirius, Pollux and Betelgeuse), there are thousands of other stars that can be viewed even without telescopic aid, and it became necessary to devise a systematic naming scheme, one that also provides some information about the stars themselves.

To this end, in 1603, the German astronomer Johann Bayer instituted a naming scheme that is based on the order of brightness of the stars within different constellations. The brightest stars are referenced with a Greek letter followed by the constellation name: for example, the brightest star in Orion is called Alpha Orionis; the second brightest is Beta Orionis; the third brightest is Gamma Orionis, and so on. Of course, the Greek letters run out before long, and other naming schemes are required. In fact, with present technology, literally millions and millions of stars and other objects have been detected, so that astronomers have now abandoned constellation-related names and have resorted to identifying individual objects with their sky coordinate positions, such as the quasar QSO 0957+1561.

Heavenly views – backyard observing

One of the things that has made astronomy so popular is the fact that anyone can do it just by going outside. Although you may not have access to world-class telescopes, there are plenty of fascinating objects to be seen using just the eye, a set of binoculars, or a small- to medium-sized telescope. There are several informative resources on backyard astronomy, which provide current news and viewing tips for observing the night sky. Here we give just a few general pointers on what to see and where to look.

Arguably the most spectacular astronomical sights are those provided by comets. Comets are occasional visitors from the outer reaches of our Solar System. As they near the Sun, the ices of the comet evaporate, leaving a trail of dust debris and gas. This material reflects the sunlight and appears as a comet's tail. Some comets pass by very infrequently, a good example being Hyakutake, which was seen in 1996 and will not return for at least 14 millennia. Other comets, however, are regular or periodic visitors. Probably the most famous is Halley's comet, which makes a trip into our neighborhood of the Solar System every 76 years. Other periodic comets include Encke, with a short period of 3.3 years, and Hale–Bopp, which was first seen in 1997 and has an estimated period of more than 2500 years.

Also spectacular, and more frequent in occurrence than comets, are the annual meteor showers. A meteor, or shooting star, is a piece of inter-

Magnitudes

The system of magnitudes used to describe the brightness of stars dates back to Hipparchus of Nicea in the 2nd century BC. Magnitudes are primarily a logarithmic scale of relative brightnesses. A logarithmic scale is one that is based on powers of 10 like scientific notation. For example, the logarithm of 10 (10^1) is 1, the logarithm of 100 (10^2) is 2, and so on. Logarithms can also be negative; for example the logarithm of 0.1 (10^{-1}) is −1, 0.01 (10^{-2}) is −2, and so on. The magnitude system operates in such a way that two stars separated by 1 magnitude will differ by a factor of 2.5 in brightness. A star that is 5 magnitudes greater than another is therefore 100 times brighter in appearance.

This system is very much like the Richter scale for measuring earthquakes. An earthquake of 7 on the Richter scale is 100 times stronger than an earthquake measuring 2 on the Richter scale. The only difference is that on the Richter scale stronger earthquakes have higher values, but for magnitudes, brighter stars have lower values. For example, the star Vega (Alpha Lyrae) has a magnitude of 0, which is 100 times brighter than a star of 5th magnitude. Magnitudes can also be negative, as for example the full Moon, which has a magnitude of −13 and is therefore 15 million times brighter than Vega. Humans can see stars with the unaided eye down to about 6th magnitude. But why do brighter stars have lower magnitudes? The logic was that the brightest stars were deemed 1st magnitude, the next brightest 2nd magnitude, and so on.

◀▶ *This mosaic of images illustrates the variety of objects that can be observed in the night sky, although not all with this kind of detail! Top left is a spectacular 35mm camera photograph of Comet Hale–Bopp as seen in late 1997 over California. Top right is the total solar eclipse of 16 February 1980. The prominent streaks are telltale signs of the faint tenuous solar corona. Middle left is a Clementine spacecraft composite image of the Moon in 1994. Middle right is a Hubble Space Telescope (HST) image of the red giant star Mira (on the right), along with its companion. Mira is also known as Omicron Ceti, and it is 700 times larger than the Sun. HST has captured (bottom left in a false-color image) the inner Galilean moon Io as it transits Jupiter. The dark spot is Io's shadow cast on to the clouds of Jupiter's atmosphere. Bottom right is an HST image of the Ring Nebula (M57). One of numerous planetary nebulae that can be seen with a modest telescope, this nebula is about 2000 light-years away and 1 light-year across. These glowing gases are actually atmospheric layers that have been ejected from the star, which can be seen as a white pinprick of light at the center of the nebula.*

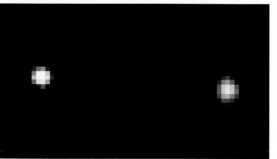

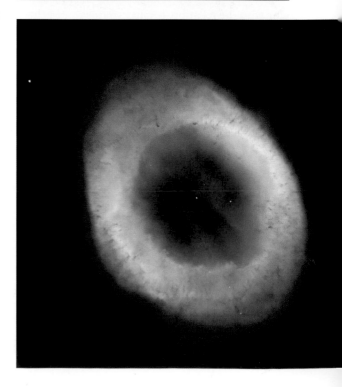

Astronomical extremities – sizing it up

This table gives examples of scales typically encountered in astronomy. The table begins with the small (a hydrogen atom) and increases in both mass and size to planets, stars, galaxies and objects at the farthest reaches of our Universe.

object	distance from Earth	mass	dimension across the object	comment
hydrogen atom	n/a	1.67×10^{-27} kg	5×10^{-11} m	The composition of the Universe is about 70% hydrogen by mass.
Earth	n/a	5.97×10^{24} kg	12,756 km	Earth is the largest of the rocky planets in the Solar System.
Sun	150 million km	2.0×10^{30} kg	1,400,000 km	The Sun is a typical star in our Galaxy.
nearest star (Proxima Centauri)	4.2 ly	$0.2\ M_{sun}$	200,000 km	A light-year is the distance traveled by light in 1 year, equal to about 1×10^{13} km.
center of the Milky Way galaxy	26,000 ly	$100 \times 10^9\ M_{sun}$	160,000 ly	The Milky Way is flattened in shape. It is 100,000 ly across but only about 1000 ly in the vertical direction.
a nearby galaxy (Large Magellanic Cloud)	170,000 ly	$10 \times 10^9\ M_{sun}$	60,000 ly	The nearest major galaxy is the Andromeda Galaxy, a near twin to our own Milky Way.
Local Group	n/a	$500 \times 10^9\ M_{sun}$	3 Mly	The Local Group is a small galaxy cluster containing about three dozen galaxies, including our own.
Virgo Cluster	60 Mly.	$1 \times 10^{15}\ M_{sun}$	20 Mly.	The Virgo Cluster is a large galaxy cluster that contains multiple thousands of galaxies.
quasars	> 1 Gly	$1 \times 10^9\ M_{sun}$ (supermassive black holes)	1000 AU (AGN)	Quasars are among the most distant and energetic objects known to man.

Footnotes: a) M_{sun} = mass of the Sun; b) AU = astronomical unit, the Earth–Sun distance; c) ly = light-year d) AGN = Active Galactic Nucleus, where supermassive black holes are believed to power quasars and related objects.

planetary debris that has been captured by the Earth's gravity; its rapid pace while falling through the atmosphere produces considerable friction, which causes the projectile to burn up, thus appearing as a shooting star. There are several annual meteor showers to observe, examples being the Perseids (August), the Orionids (October), the Leonids (November) and the Geminids (December).

Objects that are more accessible for viewing pleasure on a nightly basis are the Moon, planets, stars and nebulae (gaseous clouds in space). The first, of course, can be seen plainly with the naked eye, although not in great detail. Even a modest pair of binoculars can magnify the Moon sufficiently to make "lunar exploration" a fun activity, for sketching the terrain, identifying craters and maria, or for studying the path of the terminator (the boundary between the light and shadowed parts of the Moon).

For viewing the planets, a telescope or pair of binoculars is needed. The best views come from watching the changing phases of Venus (as

Galileo Galilei did in the 1600s), discerning features on Mars, plotting the courses of the Galilean moons as they orbit Jupiter, or marveling at the breathtaking rings of Saturn. Uranus and Neptune appear only as small blue-green spots or disks in the sky. At the very least, a telescope of 10 inches (250 mm) is required to find Pluto, which, being the smallest and farthest planet from the Sun, is only a 14th-magnitude object at its nearest approach to Earth. It is about 1 million times fainter than the brightest star, Sirius.

For those who desire a greater challenge, there are numerous glorious views to be obtained of stars and nebulae. To mention just a few, the double star Gamma Andromedae presents a nice orange and blue contrast pair, and Beta Cygni a yellow and green pair. For stars that change in brightness, a good choice is Algol (also known as Beta Persei), which is an eclipsing binary system that varies by a factor of three over the course of about three days. In an eclipsing binary, the binary orbit brings one of the stars across our viewing sight-line to the other, and when the two stars eclipse, the overall brightness is momentarily but periodically diminished (it is the same principle that operates for solar eclipses, for which the Moon passes between the Sun and Earth).

For an even greater challenge, try to see the spectacular globular cluster M13 in Hercules, or the Andromeda Galaxy (M31), the nearest major galaxy to our own Milky Way, or even the Dumbbell Nebula, a planetary nebula located in the constellation of Vulpecula, evidencing the death throes of its central star. These are just a few suggestions to whet the appetite.

Whereas the Moon, planets and prominent comets are relatively easy to identify, more challenging objects like the stars and nebulae may require aids to locate them. For these objects it is probably necessary to have sky charts that show the constellations and objects of interest. For example, Beta Cygni, a nice double star of different colors, is located in the constellation of Cygnus in the northern hemisphere. The first thing to do is find Cygnus in the sky: it is locat-

Apparent versus true stellar proximity

Although two unassociated stars may have the same apparent brightness (the brightness as seen at the Earth), the intrinsic luminosity of the two stars can be radically different. One of the stars could be relatively faint but quite near, thereby appearing fairly bright, whereas the second could be tremendously bright but at a great distance away so that it happens to have the same brightness as the first. It is the same principle as when a small torch held at a distance of only a few meters appears just as bright as the headlamp of a truck seen from hundreds of meters away. Consequently, apparent brightness is not a good gauge of actual distance nor of intrinsic luminosity.

Similarly, two stars that appear quite close together in the sky could in actuality be widely separated in distance from the Earth. To understand this effect, hold one index finger at the end of your nose, and the other at arm's length, the two can be brought very close in projection (even to appear as overlapping), yet the fingers are separated by the length of your arm. It is certainly true that many stars are binaries (consisting of two stars orbiting one another), which are, therefore, truly physically associated, but beware that apparent closeness in the sky does not always imply physical proximity.

ed east of Draco, which is itself east of Ursa Major (also known as the Plow or Big Dipper). West of Cygnus is the constellation of Lyra, and somewhat south of Cygnus is the constellation of Aquila. (There are actually two smaller constellations between Cygnus and Aquila.) The brightest star in Cygnus is called Deneb, the brightest in Lyra is called Vega, while that in Aquila is called Altair, all of which are Arabian names. These three bright stars form what is known as the Summer Triangle, which is fairly easy to locate. Having found the Summer Triangle, the constellation of Cygnus can be identified by a pattern of stars that appears as a cross, called the Northern Cross, with Deneb located at the top of the cross. The double star Beta Cygni is at the foot of the cross. In this example constellations are useful as pointers and guides that help the observer to locate interesting objects in the night sky.

The black of night

Although most of the planets and bright comets are easily seen from city areas, a darker location may be required to observe the fainter stars and nebulae. Street lights, car head lamps and other light sources all contribute to the light pollution associated with populated areas. The city lights are scattered by dust and air particles to create a luminous haze. This haze is especially evident to anyone driving in the countryside on an overcast night, when prominent cities can be identified by their reflection off the clouds. For astro-

nomical observing, light pollution serves to "wash out" night-sky objects, meaning that faint stars and other objects cannot be seen against the relatively bright background of this luminous haze. It will probably be necessary, therefore, for those interested in doing dedicated amateur astronomy to retreat to remote sites. Light pollution is one reason why world-class observatories used by professional astronomers are located in unusual areas, such as the mountains of Chile, the South Pole and space.

▼ *This is a composite picture of the Earth at night as seen from a satellite in orbit. Bright areas represent city lights, clearly revealing the light pollution associated with areas of high population density.*

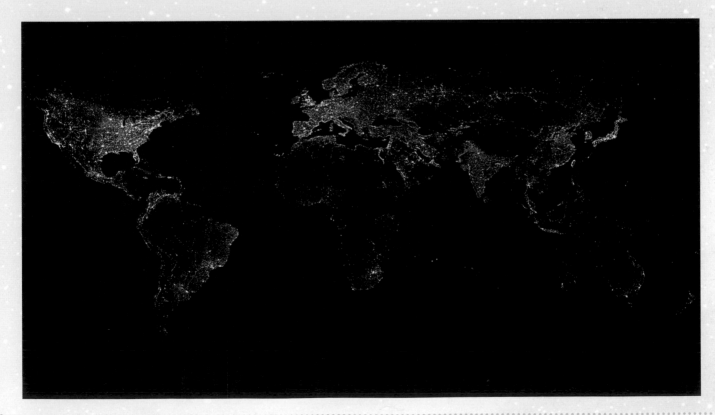

Looking through Hubble's eyes

2

Everything that has been discovered about the Universe beyond our Solar System has been discovered simply by looking; that is, by collecting and studying the often very faint light that reaches us from distant stars and galaxies. Though the human eye is a fantastic detector of light, on its own it is not adequate to meet the demands of modern astronomy. Today we need better, sharper and more detailed images than those that can be provided by the eye alone. We also need to study regions of the electromagnetic spectrum beyond visible light, such as infrared, ultraviolet and X-ray. To meet these demands, professional astronomers work with giant, powerful telescopes, either perched on isolated mountain tops or located beyond the atmosphere, in orbit around the Earth. The telescopes are fitted with an array of sophisticated light detectors and computer technology. The modern astronomical observatory is, in fact, an astrophysical laboratory.

Throughout the history of astronomy, technological advances have gone hand-in-hand with scientific progress. Our view of the Universe has become broader and much more detailed as better detectors, larger telescopes and stable platforms have been built in space. These advances are critical, since astronomy is fundamentally an observational science. To improve our understanding of the cosmos, astronomers today first collect and record light from space using a telescope. They then store the data in digital form on a computer or occasionally on photographic film. Finally, the data are analysed and interpreted using physical laws to develop a scientific theory that explains what has been observed. The conclusions reached may then be tested or refined with the help of new observations.

There are numerous professional telescopes and observatories located all over the world, producing a constant stream of valuable data and the occasional spectacular discovery. These efforts are superbly supported by dedicated amateur astronomers who methodically search the skies for comets, asteroids and variable stars using more modest instruments. However, the single astronomical facility that has attracted greatest scientific, public and media interest over the past decade is the Hubble Space Telescope. It has, quite simply, opened up a new era in astronomy.

There is no doubt that since the Hubble Space Telescope came into operation in 1990, it has provided us with some of the most dazzling and detailed views of the Universe ever seen. Many of these images are featured in this book. Its cameras have looked back at the most distant galaxies, glimpsed the eruptive lives of stars in our own Galaxy, and witnessed storms raging on planets in our Solar System. This complex observatory in orbit around the Earth has made many impressive scientific discoveries, while at the same time engaging the imagination of the public. In this chapter, we take a close look at telescopes and observatories in general, and at the unique Hubble Space Telescope in particular.

What is a telescope?

For centuries, men and women have gazed at the night sky and glimpsed the wonderful sights of astronomy using nothing more than the human eye. Meteor showers, streaking comets, wan-

► *There has been remarkable progress in the development of the telescope over almost 400 years, from the ones used by Galileo, which were made of wood and paper and had an aperture of 26 mm (1 inch), to the Hubble Space Telescope (shown here), which has an aperture of 2.4 m (94 inches) and is in orbit around the Earth.*

dering planets and fuzzy nebulae can be the rewards in a Moonless and clear night sky. But to truly appreciate the Universe beyond our Solar System, we need a better method of collecting light. The light from these very distant astronomical sources is faint when seen from Earth. To see fainter objects in the heavens, and with greater detail, it is necessary to gather more light than the human eye can manage. In the pupil of the human eye, light can only pass through an opening of a few millimeters across. The need for a much larger "bucket" to collect light led to the revolutionary use of telescopes in astronomy.

The primary purpose of a telescope, therefore, is to collect light. Professional astronomers today use telescopes that can be 10 m (33 ft) or more in diameter. The 20th century witnessed not only a growth in the number of astronomical observatories, but also great strides in the building of even larger telescopes. A bigger telescope can provide a bigger image, but this magnification is not as important as you might think. When you look at a small, blurry image and magnify it, you usually get a larger, blurry and dimmer image. Large telescopes do, however, improve the details that can be discerned in the image.

The telescope has come a long way since its invention in the early 1600s in Europe. It was Galileo Galilei, the Italian mathematician and philosopher, who, in 1609, first made the instrument famous. He used crude telescopes barely 3 cm

▶ A view of the control room of the European Southern Observatory's Very Large Telescope (VLT) in Atacama, Chile. This is the typical environment of the professional astronomer working with the most powerful telescopes.

▼ The powerful telescopes of the European Southern Observatory are located on the summit of La Silla in the Atacama Desert, Chile. The summit houses more than 15 instruments, which are used to explore skies in the southern hemisphere.

(about 1 inch) across to observe features on the Moon and the four main moons of Jupiter. His observations of the phases of Venus proved that the planet was orbiting the Sun. Over the centuries famous astronomers such as Johannes Kepler and Christiaan Huygens have used, adapted and improved the telescope.

The modern astronomical observatory

Advances in technology, computing and the manufacture of materials have led to the construction of huge modern observatories, which are in fact highly specialized laboratories. In order to decrease the adverse effects of clouds, rain and artificial lighting from bright cities, the observatories are usually located on high, isolated mountain peaks. The giant Keck telescopes, for example, have 10-m (33-ft) primary mirrors and are perched over 4000-m (13,000-ft) high on Mauna Kea in Hawaii. European astronomers regularly access a suite of powerful telescopes located at altitudes of almost 2500 m (8200 ft) in the deserts of Chile.

So who works at these isolated observatories? The stereotypical view of astronomers is perhaps of eccentric people dressed in white laboratory coats, peering through a telescope night after night looking for spectacular new sights! The research work of professional astronomers is not conducted like that at all. Rarely do they actually "look through" a telescope, but instead they sit in

control rooms and remotely steer computer-controlled telescopes. The data received are viewed as electronic images on a computer screen.

Astronomers may have access to the largest and most powerful telescopes for only one week or so in a year. They would aim, therefore, to complete a very specific preplanned sequence of observations, perhaps to find out about the chemical composition of a particular group of stars, or to survey thousands of stars in a distant galaxy. Attached to these telescopes are very sensitive instruments for gathering and investigating the light of celestial objects. This means that astronomers can observe very faint objects and can take shorter exposures so that more observations can be completed each night. Astronomers may take images at the telescope, such as that of the Crab Nebula seen on page 74; these two-dimensional views of the sky can be used to determine shapes of objects and the relationship between them. In addition, astronomers often use a very important instrument called a spectrograph (or spectrometer) to split the light from an object into its different component wavelengths (or colors). This allows a spectrum to be recorded, which can then be analysed to reveal, for example, the temperature, chemical make-up, or density of the object.

"Twinkle, twinkle little star, how I wonder what you are..."

This favorite children's poem highlights a major problem faced by astronomers, and the reason why there is a strong need to have a telescope in space. Telescopes like those located in Hawaii, Chile and the Canary Islands, for example, carry out tremendous scientific work and have produced numerous advances in astronomy. However, because they are located on Earth, their performance is restricted by the fact that the Earth's atmosphere is not very clear. Stars "twinkle" in the night sky because we are looking at them through layers of dust, pollution and even cloud. These layers move around, causing anything we can see through the atmosphere to jiggle, or twinkle. This distortion of the light interferes with astronomical observations, making images blurry – much like taking a snapshot of

Main astronomical telescopes in use today:

Refracting telescopes

These instruments use lenses to focus and produce the images. Galileo's telescopes and many of the earliest ones were of this type. The path followed by light is bent when it passes through a glass lens. If the lens is carefully shaped, it can bring rays of light to a focus, where the image or "picture" of a distant astronomical object can be examined. A second lens is placed at the eyepiece to magnify this focused image. Larger telescopes require larger and heavier lenses, but they are harder and more expensive to support and maneuver. The largest refracting telescope in the world is at Yerkes Observatory (near Chicago in the USA); it measures 1 m (40 inches) across, and its total length is more than 19 m (62 ft).

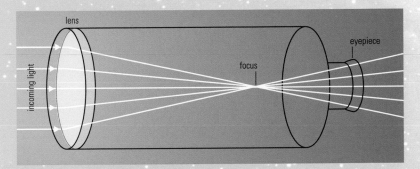

Reflecting telescopes

These telescopes use mirrors, not lenses, to produce the images. All the modern large professional telescopes in the world today are reflecting telescopes. (They are also more popular with amateur astronomers.) This type of telescope was first successfully built by Isaac Newton in 1668. A carefully shaped concave mirror (called the primary mirror) is used to bring the light to a focus. An additional (secondary) mirror is then used to reflect the astronomical image outside the main tube of the telescope. Large mirrors are easier to support than lenses, and these telescopes are generally easier to manipulate. The Hubble Space Telescope is a reflecting telescope; its primary mirror is almost $2\frac{1}{2}$ m (98 inches) across.

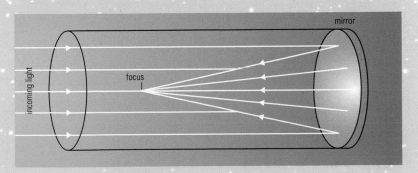

The electromagnetic spectrum

Our eyes are designed to detect a certain kind of radiation, which we call visible light. But if you pass this light through a prism, a band of colors like a rainbow comes out of the other side. Visible light is composed of all the colors of the rainbow, which can be seen in the specific order red, orange, yellow, green, blue, indigo and violet. No matter what rainbow you watch, or what prism you use, the colors are always in this order. However, the colors that make up visible (or white) light are just part of what is called the electromagnetic spectrum. There is radiation beyond the violet color we can see, and beyond the red on the other side of the rainbow. We cannot see this additional radiation, but we can directly experience some of it, as with sunburn for example.

All of the other components of the spectrum are important in astronomy. The most energetic radiation is in the form of gamma rays. First discovered in studies of radioactive atoms, gamma radiation is generated deep in the cores of stars.

More familiar, perhaps, are X-rays, which are used in hospitals to look at bones in the human body, for example. The X-rays are more energetic than visible light and are able to pass through soft body tissues but not through bones; this creates shadows which can be imaged. Very hot gases in the Universe, at around a million degrees or more, can produce X-rays.

Ultraviolet radiation simply means "higher energy than violet." We know the Sun is a source of ultraviolet radiation, since its UV rays can give us sunburn. Our eyes do not see infrared radiation either, but we can feel it as heat on our skin. Infrared radiation is intermediate between visible light and radio waves. All the electromagnetic radiation beyond infrared is called radio waves, and this includes familiar categories such as microwaves, radar waves, FM radio and television waves. In astronomy, the radio region is especially useful for mapping our Galaxy.

◄ *The Space Shuttle Discovery is launched on 24 April 1990 from the Kennedy Space Center in Florida, USA. It carried five astronauts and the 12,500 kg load of the Hubble Space Telescope on a five day mission.*

something in motion. This presents a major problem when trying to study the most exquisite details or the faintest objects.

The problem can be solved by rising above the atmosphere and observing from space. A telescope above the Earth's atmosphere – like the Hubble Space Telescope – can see things much more clearly, thus offering exciting new opportunities to learn about how our Universe works.

Another very important reason for placing telescopes in orbit above the Earth is so that we can study radiation from stars and galaxies in other parts of the electromagnetic spectrum, which cannot penetrate the Earth's atmosphere. Most of the electromagnetic radiation from space, like harmful X-rays and ultraviolet (UV) rays, do not reach the ground. But this kind of light is of great interest to astronomers, as it often provides vital information that can improve our understanding of an object, such as an extremely active galaxy or the hottest stars. Once again, it is only from space that this light can be collected, using dedicated orbiting satellites. These satellites are not optical telescopes, like most of those on the ground, but instead they house modest telescopes and detectors tuned to see only in the infrared, ultraviolet or X-ray. They are usually controlled from one or two ground-stations on Earth, which beam up computerized instructions to point the telescopes and carry out the observations.

The Hubble Space Telescope

The Hubble Space Telescope is the first large telescope to be placed in space that can see in visible light. Its view is not distorted by the hazy atmosphere of the Earth. It also has instruments on board that allow ultraviolet and infrared radiation to be detected. It was built by and is operated as a cooperative program between the National Aeronautics and Space Administration (NASA) of the USA and the European Space Agency (ESA).

The idea of a dedicated space-based observatory was first discussed in the 1940s, but the Hubble Space Telescope itself was designed and built in the 1970s and 1980s and did not become operational until 1990. It was a major and very

▶ *The Hubble Space Telescope is shown here on the Space Shuttle Discovery's remote manipulator arm in a photograph taken on 25 April 1990. Its solar panels and antennae are about to be deployed before it is released above the Earth.*

costly undertaking. So far, the total cumulative cost of the Hubble project is about US$7 billion. It has an annual budget of around US$250 million to maintain operations, carry out servicing, develop new technologies, and investigate the scientific observations.

The Hubble Space Telescope is a reflecting telescope, with a primary mirror that is almost 2½ m (98 inches) across. It has a variety of scientific instruments, which can be upgraded or replaced by visiting Space Shuttle astronauts. The Telescope weighs 12½ tons and is 13 m (43 ft) in length, which is about the size of a city bus. Power to the onboard computer and scientific instruments is provided by two solar panels, which are 12 m (39 ft) long by 2½ m (8 ft) across. The Telescope has a planned 20-year operations lifetime, with several servicing missions scheduled for replace-

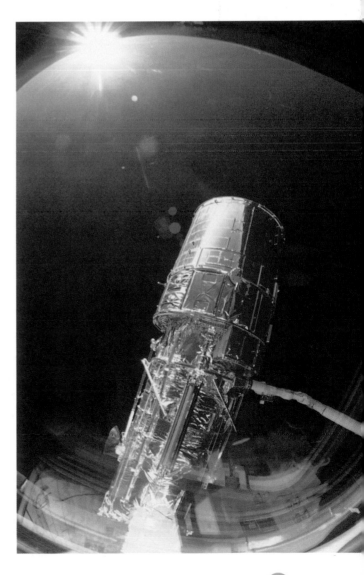

ment and repair of its hardware. It was launched on 24 April 1990 on the Space Shuttle *Discovery* (STS-31 mission), and placed in low-Earth orbit, 613 km (381 miles) above the Earth. Traveling at almost 30,000 km/h (19,000 mph), it circles the Earth once every 90 minutes. The Telescope is in almost continuous operation and each day transmits enough data to fill an encyclopedia.

So how do scientists actually get to use the Hubble Space Telescope? Each year, astronomers working at universities and scientific institutes around the world submit written proposals to the Space Telescope Science Institute (STScI) at Baltimore, Maryland, USA. The staff at this institute are the liaison between the Hubble Space Telescope and the international community of professional (and amateur) astronomers. Panels of scientists judge the proposals on scientific merit, and select those that address the most important astronomical questions and represent the best use of the Telescope. Typically, about 300 proposals are selected each year out of more than a thousand. The approved proposals are then scheduled into the overall observing program. Once an observation has been carried out, the computers on the Telescope convert the image or data into a format that can be transmitted down to Earth via communication satellites. The information is reconverted into the desired images and stored on computers. The astronomers can then study and analyse the data at their home universities and institutions.

An interesting question is how does the Hubble Space Telescope point itself to different directions in the sky to observe different objects? There are no rockets on the Telescope since the fumes would

Edwin P. Hubble

The Hubble Space Telescope is named after Edwin Powell Hubble, an American astronomer who transformed our understanding of galaxies and the nature of the Universe. Edwin Hubble was born on 20 November 1889 in Missouri, USA. As a school boy he enjoyed books and writing, and excelled in a range of sports. He won a scholarship to study at the University of Chicago, from where he received his Bachelor degree in 1910. He went on to read Roman and English Law at Queen's College, Oxford, UK. His true passion, however, was astronomy, and so he returned in 1914 to the University of Chicago to work for a postgraduate degree in astronomy.

After World War I, Edwin Hubble worked at Mount Wilson Observatory in California. He later greatly assisted in the design of the 200-inch (5-m) Hale Telescope at Mount Palomar Observatory, also in California. Edwin Hubble made several revolutionary contributions to astronomy: in particular he proved the existence of other galaxies beyond our Milky Way. He then went on to classify galaxies according to their appearance. Most significantly, his measurements of galaxies in the 1920s, with the American astronomer Milton Humason, led to proof that the Universe was expanding. He discovered the relationship between a galaxy's distance and the speed with which it was moving away from us, determined by the Hubble Constant. These studies, which had a great impact in determining the extent of the Universe, are explained further in Chapter 6.

Edwin Hubble's remarkable work was fittingly recognized by the astronomical community in 1983, 30 years after his death, when the Large Space Telescope was renamed in his honor.

▲ *Edwin Hubble is seen working at the 48-inch Schmidt Photographic Telescope (more correctly known as the Schmidt Camera) at Mount Palomar Observatory. He is preparing for the National Geographic Sky Survey, a project that provided the first ever definitive photo atlas of the skies.*

▼ Shown here is a scene
from the first servicing
mission to the Hubble
Space Telescope. A
remarkable series of
spacewalks was completed
over five days to improve
the vision of the telescope
and to replace instruments.
Astronaut Story Musgrave
is seen anchored to the
end of the Shuttle's
Remote Manipulator arm,
while Astronaut Jeffrey
Hoffman is at the bottom
of the image.

soon make its mirrors dirty. Instead, the Telescope uses some very basic principles of physics to help it turn. It has four reaction wheels, which can spin freely. When the Telescope needs to turn, a command is given for one or more wheels to spin. Now, according to Newton's laws of motion, every action has an opposite and equal reaction. This means that as the reaction wheels spin in one direction, the Telescope responds by rotating in the opposite direction. The Telescope can therefore be steered toward any location in the sky by using its combination of four reaction wheels.

Hubble, toil and trouble

Astronomers anticipated that the Hubble Space Telescope would provide very detailed images, resolving subtle features ten times better than any telescope situated on Earth. Within a few months

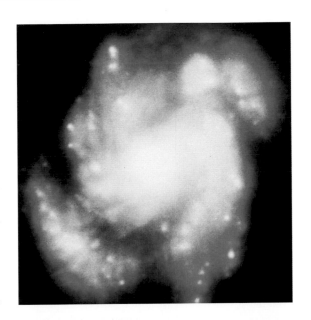

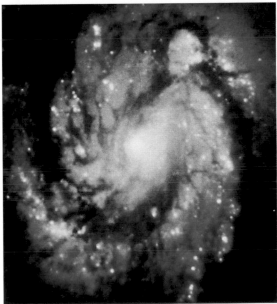

▲ These images show the dramatic improvement in the Hubble Space Telescope's vision after it had been serviced by the NASA astronauts in 1993. Both pictures are of a galaxy called M100, which is tens of millions of light-years away from us. At the top is the image taken in November 1993 just a few days prior to the servicing mission. The effects of the faulty mirror are clear as much of the detail is blurred. The image at the bottom was taken on 31 December 1993 after the optics had been corrected. The Hubble Space Telescope could now for the first time reveal exquisite fine details in the galaxy. Structures only about 30 light-years across are resolved in the improved picture.

of launch, however, a flaw was discovered in the Telescope's main (primary) mirror which greatly reduced its ability to focus. Though the problem was perhaps not as terrible as reported in the popular media at the time, it was, nevertheless, a serious blow for the scientists.

The primary mirror of the Hubble Space Telescope is one of the most finely polished mirrors ever made. However, during the construction of the mirror there had been an error in the testing equipment used to shape and polish the incredibly smooth surface. This means that the mirror on the Telescope has a slightly imperfect shape; it is too flat by about 1/50th of the width of a human hair. This tiny error, though, meant that instead of being focused to a sharp point, light collected by the mirror was spread over a large fuzzy area. Images of planets, stars and galaxies were thus slightly blurred. Though the imperfect Telescope was still making outstanding discoveries, scientists knew that it had a far greater potential.

It had always been the plan to visit the Hubble Space Telescope on Space Shuttle missions every three years or so to replace worn-out equipment and to install more technologically advanced instruments. This would ensure that the Telescope remained an outstanding facility for its planned 20-year lifetime. It became necessary, however, to use the first servicing mission to restore the Telescope's focus.

On 2 December 1993, seven astronauts were launched on Space Shuttle *Endeavour* on an amazing 11-day mission to repair the Hubble Space Telescope. Their main task was to install a refrigerator-sized box called COSTAR (Corrective Optics Space Telescope Axial Replacement), which would act like a pair of spectacles for the Telescope and greatly improve its focus. The Telescope has special built-on fixtures to allow astronauts to capture it with the Shuttle's robotic arm and lock it in the payload bay. There are also hand-holds to help astronauts

Ten amazing facts about the Hubble Space Telescope

1 If the Hubble Space Telescope looked down at you and there were no distorting effects from the Earth's atmosphere, it would be able to see your eye sockets, mouth and ears.

2 The height of the Telescope's orbit above the Earth is equivalent to the distance between London (England) and Frankfurt (Germany).

3 The Telescope was lifted into orbit by the Space Shuttle, the three main engines of which develop a power equivalent to 37 million horsepower.

4 The Telescope's 12-m-long solar panels can generate 2000 watts of electricity, which is enough to power more than 30 household light bulbs.

5 If the ordinary lenses in a pair of spectacles were scaled to the size of Europe, the typical imperfections of the surface of the lenses would be the size of skyscrapers. By contrast, on this scale the ultrasmooth mirror of the Telescope would have flaws that were barely 5 cm (3 inches) high.

6 The Telescope has measured the speed of gas swirling around a massive black hole in the center of a galaxy at over 1.5 million km/h (930,000 mph).

7 The images obtained by the Telescope travel almost 150,000 km (93,000 miles) between several satellite and ground links before they finally reach the computers in Baltimore, Maryland, USA.

8 Each day the Telescope transmits enough data to fill 2000 computer floppy disks.

9 To help astronauts service it, the Telescope is fitted with 31 foot restraints and almost 70 m (230 ft) of handrails.

10 To point the Telescope and keep it locked on to a distant star is like holding the light from a laser gun on a small coin placed almost 700 km (435 miles) away.

move around the Telescope while they work in a weightless environment. Doors open up all around the Telescope, giving astronauts convenient access to the many electronics and scientific instruments.

Remarkably, the astronauts accomplished everything that they had planned to do in a record mission of five spacewalks. They even upgraded computer components and replaced faulty solar panels. The mission was a tremendous success and rates as a great triumph of space engineering. Through incredible ingenuity the Hubble Space Telescope had been transformed from an under-performing instrument into the most valuable tool for unravelling the mysteries of the Universe.

Making awesome pictures

Let us look briefly at how the Hubble Space Telescope pictures are actually generated. Remember, though, that only a relatively small fraction of the Telescope's time is spent taking pictures, with the majority of its time dedicated to the spectrographs, which split the light to make spectra.

When in imaging mode, the Hubble Space Telescope directs light from a star or galaxy to special cameras that can record the information. These cameras are electronic sensors similar to those found in camcorders, although they are, of course, much more sensitive to light since most objects in space are very faint. Many of the Telescope images discussed in this book were taken using the Wide Field and Planetary Camera. It records its images on four devices known as CCDs (Charge Coupled Devices), instead of film. The four CCDs are arranged to cover a bigger sky area or field than would be possible with just one CCD alone. The data from the cameras are then transmitted to Earth as numbers, the values representing brightness measured by each detector in the camera. Generally, the region of sky recorded at one time is not very large, being, for example, much less than the diameter of the full Moon.

The cameras on the Hubble Space Telescope are not color ones, and the pictures are recorded on Earth only in black and white. Astronomers can then generate color images from the black-and-white ones in two basic

▲ *A last majestic view of the newly refurbished Hubble Space Telescope. It is ready to probe even further into the mysteries of the Universe.*

ways. First, a color may be assigned to each number transmitted by the Telescope. These colors are arbitrary and the images are usually called false color. A color palate on a computer can also be used to highlight different parts of an image. Alternatively, more natural color images can be made by combining several black-and-white pictures that have been viewed through different colored filters. This is similar to the way in which color film photography works in our home cameras.

Expanding the vision

The Hubble Space Telescope is one of the most advanced and complex space observatories ever built. Since 1990 it has been providing clues to some of the most perplexing questions in astronomy: What is the age of the Universe? How do black holes form? What is the nature of the objects in the Universe? It has revealed astounding new findings, such as conclusive evidence for

the existence of massive black holes, some of the deepest views of the early history of the Universe, and exquisite snapshots of the birth and death of stars. But the Telescope was designed so that it would remain at the forefront of scientific discovery for its entire 20-year life span, up to the year 2010. This objective is being achieved by servicing the Telescope approximately every three years in order to replace scientific instruments with more technologically advanced ones.

In February 1997 astronauts aboard Space Shuttle *Discovery* carried out the second servicing mission. The astronauts installed two new devices to increase the capability and range of the Hubble Space Telescope. The first device, called the Near Infrared Camera and Multi-Object Spectrometer, samples a region of the electromagnetic spectrum never previously seen from space. The detectors of this instrument must be kept very cold to work best and are chilled to less than −215°C (−355°F) using solid nitrogen. The other instrument, the

▼ *Astronauts Mark Lee (right) and Steve Smith are on a spacewalk to install new instruments on the Hubble Space Telescope during the second servicing mission, which was carried out by the Space Shuttle* Discovery *in February 1997.*

Space Telescope Imaging Spectrograph, divides light into its component colors. This instrument can gather information from several hundred points along an astronomical object, recording, for example, many stars within a galaxy in a single exposure. This ability greatly improves the Hubble Space Telescope's efficiency and speed.

A follow-up mission to the Telescope in December 1999 was used to install six new gyroscopes, a new computer and new batteries. The next Shuttle visit was in March 2002. On this trip the astronauts installed an advanced camera, a new system for keeping the spectrometer cold, and a new set of solar panels. A fourth servicing mission is planned for 2004, during which a new ultraviolet spectrometer and another powerful camera will be installed.

The next generation

The Hubble Space Telescope is expected to operate until around 2010. In spite of the tremendous new discoveries made and the vast increase in our knowledge of the Universe, many fundamental and very important questions remain to be answered. To tackle these issues, the National Aeronautics and Space Administration (NASA) plans to launch in 2009 a revolutionary new kind of space observatory called the Next Generation Space Telescope. This exciting and ambitious project will open up a new window to the Universe. It will provide, among other things, an improved understanding of how stars and planets form, and the origins of galaxies. It will doubtless provide many major new discoveries over the next two decades.

The Next Generation Space Telescope is planned to have a 6½-m (21-ft) primary mirror, giving it light-gathering capabilities comparable to some of the largest ground-based telescopes. It will be capable of detecting the full range of infrared radiation and will be able to see objects 400 times fainter than those that can currently be studied by giant telescopes situated on Earth. Many secrets about the birth of stars, planetary systems and galaxies can be unlocked in infrared light, which penetrates the dust clouds in space that block visible light. The budget for the con-

struction of the new telescope is US$500 million, which is less than a quarter of the cost of the Hubble Space Telescope. This smaller cost is possible because the new telescope will be a quarter of the weight of the Hubble Telescope and will not need servicing by astronauts. The Next Generation Space Telescope will be an international partnership between several countries involved in scientific and technological collaboration.

This book presents some of the most breathtaking images taken by the Hubble Space Telescope of dying stars, giant nebulae of gas and dust, regions of prolific star formation, and furiously active galaxies. We can, however, only begin to imagine the awesome views that will be captured by the even more powerful Next Generation Space Telescope. There is no doubt that our exploration of the Universe will soon reach further than ever thought possible.

▲ *Shown here is an artist's view of one of the designs being considered for the Next Generation Space Telescope. Studying the way in which galaxies and stars are born and evolve will be at the heart of this exciting new mission.*

Our Solar System and other worlds

3

The Space Age has had a great impact on modern science, and nowhere has this been more significant than in the exploration of our Solar System. Since 1959 robotic spacecraft have taken us on an incredible journey of discovery. They have visited and photographed all the planets except Pluto, as well as dozens of moons, four ring systems, and even some comets and asteroids. Some of the space probes have landed on the surfaces of Venus, Mars and Earth's moon, and one has even plunged into the gaseous depths of Jupiter. Humans have set foot on their Moon and returned to Earth with samples of its surface soil.

Most of the objects in our Solar System appear as little more than tiny, featureless images through telescopes on Earth. Spacecraft explorations have transformed them into real worlds, each with remarkable properties and unique beauty. The Solar System is rich in wonders and mystery. We are now beginning to understand how the Sun's "family" came into being and how its parts have evolved over billions of years. Today, using complex telescopes like the Hubble Space Telescope, astronomers are able to study the atmospheres and changing weather patterns of other planets over periods of months and years. Perhaps most remarkable and exciting of all, we are now beginning to gather information about giant planets orbiting other stars. All these advances have provided us with a richer outlook on our own home in space.

◀▲ Spacecraft exploration has provided us with the most detailed views of the Solar System, and has transformed our understanding of the origin and evolution of the planets. The five exciting missions highlighted here are (moving counterclockwise from left): Mariner 10, which coasted over Mercury between 1974

An overview

The term Solar System refers to the Sun (our star) and all the objects that travel around it. It is completely dominated by the Sun, which contains more than 99% of all the matter in the Solar System and is the principal source of heat and light. Because the Sun is so massive, its gravity pulls on everything in the Solar System and makes the Earth and other planets orbit the Sun in the manner that they do. The components of the Solar System can be divided into planets (the larger bodies orbiting the Sun), their satellites (or moons, which are objects orbiting the planets), asteroids (small rocky objects orbiting the Sun), and comets (small icy objects with very elongated, not circular, orbits). Bodies like the planets and their moons do not generate their own light, but instead reflect the light of the Sun.

and 1975; Venera 13 of the former Soviet Union, which landed on the surface of Venus in 1982; Apollo 17, which was (in 1972) the last of NASA's very successful Moon landing programs; the Sojourner vehicle, which roamed on the surface of Mars during July 1997; and Voyager (1 and 2), which sailed passed the outer giant planets Jupiter, Saturn, Uranus and Neptune between 1979 and 1989.

▲ *These photographs of the nine planets in our Solar System were taken by various spacecraft and telescopes. The planets are shown roughly to scale of their sizes (but not in their distances apart). The limb of the Sun is shown on the left. The four terrestrial planets (Mercury, Venus, Earth and Mars) are much smaller and denser than the giant gas planets (Jupiter, Saturn, Uranus and Neptune).*

What's in a name?

The planets in our Solar System are named after figures from Greek and Roman mythology, following a tradition from when the language of science was Greek and, later, Latin.

Mercury was so called because it moves rapidly around the Sun, just as the messenger of the gods, who had wings on his heels to carry him swiftly from Earth to the heavens.

Venus was associated with the goddess of love and creativity, since it shines so brilliantly in the early morning or late evening skies.

Earth is the only planet not named after a mythological Greek or Roman character. Its name comes from old English and Germanic languages. There are hundreds of different names for the planet in other languages.

Mars looks red in the night sky, suggesting the fire of battle. It was thus named after the god of war.

Jupiter is a giant planet that moves steadily across the sky, just as the king of gods, who rules with a steady hand.

Saturn is named after the Roman god of agriculture, who was concerned with the sowing of seeds. He is known as Cronus in Greek mythology. Cronus was the father of Jupiter.

The planets out to Saturn can be seen with the unaided eye. The rest of the planets in our Solar System were discovered only after telescopes extended the vision of astronomers.

Uranus was discovered in 1781 by William Herschel. It is named after the ancient Greek god of the heavens. The earliest supreme god, he was the father of Cronus (or Saturn).

Neptune was discovered in 1846 by Johann Galle and Heinrich D'Arrest after predictions made by John Couch Adams and Urbain Le Verrier. Its name is from the god of the sea, who was the brother of Jupiter.

Pluto was discovered in 1930 by Clyde Tombaugh, an astronomer working at the Lowell Observatory, USA. The popular press at that time encouraged the public to offer suggestions for naming the planet. Out of these, staff at Lowell Observatory chose Pluto, god of the land of the dead, which seemed appropriate for the coldest and most distant planet.

In increasing distance from the Sun, the planets in our Solar System are Mercury, Venus, Earth, Mars, Jupiter, Saturn, Uranus, Neptune and Pluto. The path of Pluto around the Sun is peculiar, however, and it sometimes crosses inside the orbit of Neptune. From 1979–99, for example, Pluto was in fact the eighth planet from the Sun. The planets all revolve around the Sun in the same direction, which is counterclockwise when viewed from the Sun's north pole. With the exception of Pluto, the planets' orbits around the Sun are nearly circular and are in a flat plane called the ecliptic; they resemble tracks on a giant dinner plate. Each of the planets also rotates about an axis running through it.

The Solar System is enormous when compared to the sizes that we are familiar with on Earth. The distance from the Sun to Pluto, for example, is almost 15,000 times greater than the distance from the Earth to its moon. Yet on astronomical scales the entire Solar System lies very close to its parent Sun. Light from the Sun, which travels at 300,000 km/s (190,000 miles/s), takes about $5\frac{1}{2}$ hours to reach Pluto but only 8 minutes to reach the Earth. In comparison, it takes several years for sunlight to reach the nearest stars.

The nine planets can be classified or grouped in several different ways. By size: the *small planets* are Mercury, Venus, Earth, Mars and Pluto; the *giant planets* are Jupiter, Saturn, Uranus and Neptune. By position from the Sun: the *inner planets* are Mercury, Venus, Earth and Mars; the *outer planets* are Jupiter, Saturn, Uranus, Neptune and Pluto. By composition: the *terrestrial* or *rocky planets* are Mercury, Venus, Earth and Mars; the *gas planets* are Jupiter, Saturn, Uranus and Neptune.

The terrestrial planets have solid surfaces and are composed mainly of rock and metal. They have no rings around them and only a few satellites (or moons). The gas planets are gigantic in comparison with Earth. They are made up mostly of hydrogen and helium gas, with relatively small solid cores at the center. All the gas planets have rings around them and lots of satellites. Tiny Pluto is very much like some of the satellites of the gas planets.

The scale of things

One way to imagine how the sizes of the planets in our Solar System compare with each other is to make up scale models. Imagine, for example, that the whole Solar System is reduced in size by a factor of 1 billion. The Earth would be about 1.5 cm (half an inch) across – the size of a grape. The Sun is then a ball 1.5 m (5 ft) across – about the height of a man or woman. In this model the Sun (or ball) would be placed 150 m (500 ft) away from the grape (Earth). The giant planet Jupiter is 15 cm (6 inches) across (a large grapefruit) and 780 m (half a mile) away from the Sun. Saturn, the size of an orange in this scale model, is almost 1.5 km (nearly a mile) away. Pluto would be a pinhead placed almost 6 km (3.5 miles) away from the Sun. On this scale a human would be the size of an atom. The nearest star from the Sun, Proxima Centauri, would be over 40,000 km (24,500 miles) away, which is as much as a complete trip around the surface of Earth. But remember we have reduced the size of the Solar System here by 1 billion: space is actually rather empty.

Another telling scale model is to imagine the Earth as the size of a beach ball (about 40 cm/16 inches across). The Moon would then be a softball (about 10 cm/4 inches across) placed about 12 m (40 ft) away. Now the Space Shuttle in orbit around the Earth would be only slightly more than 1 cm (size of a thumb nail) above the beach ball. Everything we know currently that is alive in the Universe lives on the "surface" of the beach ball, below this 1 cm height.

Where did our planets and Solar System come from?

Our Solar System was born about 4.5 billion years ago out of a gigantic cloud of dust and gas, called a nebula. This dust and gas is the raw material for making stars and planets. Some sort of disturbance, perhaps a collision with another nebula or a shock wave from a nearby exploding star, caused "our" nebula to start collapsing inward under the pull of its own gravity. Any such cloud that is rotating or spinning even very slowly will tend to form a flat rotating disk as it contracts. Most of the material works its way to the center of this disk, where it becomes tightly or densely packed. This central region eventually gets very hot, and this is where our star, the Sun, was born.

The great cloud of material in the center got hotter and hotter as it was squeezed by gravity. The temperature at the heart of the young Sun eventually reached almost 14 million °C (25 million °F),

which is hot enough for nuclear fusion reactions to occur. At this point the Sun became truly a star and began to shine. Our Sun is a star just like many of the stars that twinkle in our night sky, but because the Sun is much closer to Earth than any other star, it appears very large and bright. Without the Sun's light and heat there would be no life on Earth.

The Sun is a gigantic ball of very hot gases, mostly hydrogen and helium. This awesome nuclear power plant fuses hydrogen to make helium, and in doing so generates tremendous energy. In fact, the Sun converts 6 million tons of hydrogen into helium every second, and it is this process that makes the Sun glow. There is enough hydrogen in the Sun's core to keep our star shining for another 5 billion years. We will see later in this book what fate awaits the Sun and the Solar System once this nuclear fuel runs out.

Only a small fraction of the material from the original nebula remained in the flat disk around the newly forming star. Close to the Sun the material was dominated by rock and metal, which then clumped together into even larger bodies, eventually forming the rocky inner planets Mercury, Venus, Earth and Mars. By con-

SUN – Fact Box

Diameter: 1.4 million km (870,000 miles), which is
    109 Earth diameters
Mass: 330,000 times the mass of the Earth
Time to rotate (at equator): 25 Earth days
Temperature in central core: 14 million °C
    (25 million °F)
Temperature at surface: 5500°C (9930°F)
Distance from center of our Galaxy: 26,000 light-
    years
Age: 4.5 billion years

▲ An artist's impression of the Solar System in the process of forming about 4.5 billion years ago. The solar nebula collapsed under gravity to form the young Sun in the center, surrounded by a rotating disk of gas and dust. Material in the disk eventually collected to form the planets.

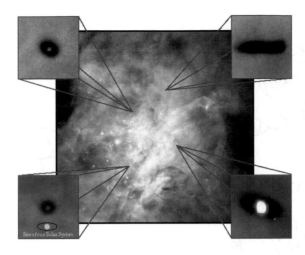

► The Hubble Space Telescope captured this view of infant solar systems forming in the Orion Nebula, which is located almost 1500 light-years from Earth. The image in the center is a close-up of the nebula itself, which contains vast clouds of gas and dust. These are the raw materials for making stars. The pictures around the nebula show brand new stars forming. In each case the infant star is still surrounded by a flattened disk. Over millions of years the disks of dust may collect into a few larger fragments to form planets. The circle and yellow dot in the lower left inset show the size of our Solar System for comparison.

trast, the outer regions of the Solar System (far away from the Sun) were chilly enough for ices and lightweight gases like hydrogen and helium to remain. Much of this material, plus some rock and metal, collected together to make up the central cores of the giant planets Jupiter, Saturn, Uranus and Neptune.

The giant planets began like the inner planets, having gathered material into solid cores. But these cores grew quickly in mass, increased their gravitational influence, and swept up much of the leftovers of the nebula, thus growing larger still. Their cores became massive enough to capture and retain gas from the nebula, which is why the giant planets are today almost entirely composed of gases. Jupiter and Saturn contain the largest amounts of hydrogen and helium. Uranus and Neptune formed in even colder regions of the Solar System, and they also contain ices of frozen water, ammonia, methane and carbon monoxide.

The planets acquired most of their moons (or satellites) at the time when they themselves were forming. Near the powerful Sun, heat, radiation and the effects of gravity prevented moons forming. Mercury and Venus, for example, do not have any moons. Moons formed much more easily farther from the Sun. Pluto, the most remote planet in our Solar System, is very much like some of the small icy and rocky moons of the giant outer planets. Its orbit is also quite different from those of the other planets. Indeed, many scientists do not consider Pluto to be a planet at all.

There are also billions of small icy bodies in the deepest outer parts of the Solar System (more than 1000 billion km/600 billion miles away). They make up a vast shell called the Oort cloud. Disturbances, such as collisions, in this "cloud" can cause icy rocks to hurtle toward the Sun in long, elongated orbits. These bodies eventually become comets, developing glowing tails of gas and dust as they approach the hot Sun.

Other leftovers from the formation of the Sun and planets are seen as pitted irregular-shaped rocks called asteroids. They are typically only a few kilometers in size, but there are millions scattered throughout the Solar System. Many of them live in the asteroid belt, which is a zone between the orbits of Mars and Jupiter. These asteroids never collected together to form a planet because the strong gravity of the nearby giant planet Jupiter kept pulling them apart.

Nine different worlds, and more

The nine planets in our Solar System are notable for their individual and widely differing properties. Each has its own unanticipated beauty and mystery. Let us take a closer look at them, and also at the nature of asteroids and comets.

Mercury – at the edge of the Sun

Mercury is the planet closest to the Sun. Of the nine planets in our Solar System only Pluto is smaller than Mercury. It is about the size of Earth's moon, and it even looks very much like our moon. Its surface is scarred by ancient craters, which were made when large rocky objects crashed into Mercury billions of years ago. These objects were part of the debris that was flying around when the Solar System was forming and settling down.

MERCURY – Fact Box

Average distance from the Sun: 58 million km (36 million miles)
Diameter: 4880 km (3032 miles)
Mass: 0.06 times the Earth's mass
Time to orbit the Sun: 88 Earth days
Rotation time (a day): 59 Earth days
Number of moons: none

Mercury is a dense, rocky planet, with a core made of iron and nickel. In its orbit around the Sun, the planet travels at almost 50 km/s (31 miles/s), which is faster than any other planet.

Mercury is so close to the Sun that midday temperatures can soar to 370°C (700°F); its surface can get hot enough to melt tin. But the planet has no substantial atmosphere to act like a blanket and hold this heat in. Night-time temperatures, therefore, drop spectacularly to almost –185°C (–300°F). These great extremes in temperature can cause cracks in the surface of the planet. Strangely, despite this enormous heat, astronomers have found that Mercury has small ice caps at its north and south poles. Scientists think they were formed when icy comets crashed into deep craters at the poles. The floors of these craters are permanently in shadow, and so the Sun's rays cannot melt the ice caps.

Mercury is difficult to view from Earth, since it never strays far from the brilliant glare of the Sun. Sometimes it is visible briefly before sunrise or after sunset, when it looks like a bright star shining just above the horizon. A new spacecraft mission, called Bepi Colombo, is scheduled for launch in 2009 to explore this planet in detail.

What would it be like to visit?

Only one spacecraft, Mariner 10, has ever gone near Mercury. It made three passes by the planet between 1974 and 1975, taking photographs and scientific readings. Mercury has no air to breathe and no water. Nothing can grow on its seethingly hot day side, nor on its excruciatingly cold night side. To live on Mercury you would need an incredibly bulky and strongly shielded spacesuit, but you would feel light because the gravity on Mercury's surface is only one-third that of Earth. You would be walking on a hard and rocky surface, splattered with numerous craters. The terrain would be broken up by huge cliffs, rising a kilometer or more above the ground. On Mercury, the Sun in the sky would look almost three times as large as it does from Earth.

▼ *Caloris basin on Mercury can be seen in the left half of this image. This major feature on the surface of the planet was created when a giant asteroid-like body hit Mercury about 3.5 billion years ago. The basin is 1300 km (800 miles) across, and its rim is marked by mountains up to 2 km (over a mile) high.*

▲ *This mosaic picture of Mercury was made from several images recorded by the Mariner 10 spacecraft as it approached and receded from the Sun-scorched planet. Most of the pictures were taken from a distance of 200,000 km (124,000 miles) from Mercury. (A strip in the upper right remains blank and was not viewed by Mariner 10.) The planet is covered with thousands of craters and resembles Earth's moon in appearance.*

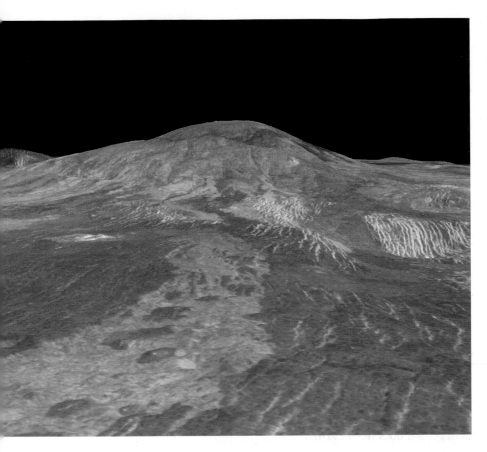

Venus – cloud-covered inferno

Venus is sometimes called Earth's sister planet, because their sizes and masses are almost the same. But Venus, the second planet from the Sun, is the hottest world in the Solar System. The temperature on its surface can reach 480°C (896°F), which is enough to melt lead. The reason for Venus' incredible heat is that it is totally shrouded under a thick blanket of clouds, containing carbon dioxide and sulfuric acid. This atmosphere acts

VENUS – Fact Box

Average distance from the Sun: 108 million km (67 million miles)
Diameter: 12,102 km (7520 miles)
Mass: 0.82 times the Earth's mass
Time to orbit the Sun: 225 Earth days
Rotation time (a day): 243 Earth days
Number of moons: none

▲ This 3-D view of the Sif Mons volcano on Venus was computer generated using Magellan radar images. Sif Mons is 2 km (over a mile) high, and has a base 300 km (185 miles) across. Most of the volcanoes on Venus have probably been inactive for hundreds of millions of years.

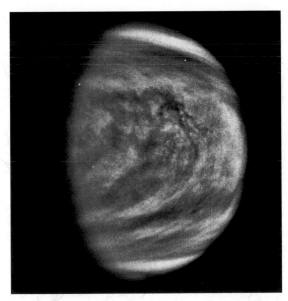

▲ The clouds enshrouding Venus prevent any view of its surface from Earth, or as in this case (left) by the approaching Galileo spacecraft, when it was 2.5 million km (1.6 million miles) from the planet.

(Galileo was actually on its course to Jupiter at this moment, but it needed a slingshot-type boost from the gravity of Venus.) The clouds at the equator, which is across the center of the planet in this image, are

moving to the left (west) at about 400 km/h (250 mph). The image on the right is a false-color global view of the surface of Venus' northern hemisphere, orientated such that the north pole is at the center.

The image was made using radar mapping by the Magellan spacecraft. Maxwell Montes, which is an 11-km-high (7-mile-high) mountain on Venus, is the bright region below the center of the image.

▲ *The Venusian surface was photographed by the Soviet Venera 13 lander in March 1982. It survived on the planet for 127 minutes. This wide-angled photograph shows a rocky surface, illuminated orange by sunlight passing through Venus' thick and cloudy atmosphere. The triangular teeth at the base of the spacecraft are 5 cm (2 inches) apart.*

like a giant greenhouse, trapping in the heat. The effect is very similar to what happens inside a car on a hot, sunny day. Sunlight comes through closed windows and warms the inside of the car, but the heat cannot escape. In the case of Venus, this heat has been building up for billions of years.

The unbroken cloud cover around Venus prevents us from viewing its surface directly from Earth. Beginning in 1975, the then Soviet Union landed several Venera spacecraft on the planet. They typically survived for about an hour, sending back pictures of a rocky surface. The United States and Soviet Union also used radar images to map the surface of Venus. From 1990–92, the Magellan spacecraft was used to radar map almost the entire planet. Scientists discovered mountains on Venus that are taller than any on Earth, and valleys that are larger and deeper than the Grand Canyon in the United States. Venus' surface has also been shaped by what were once very active volcanoes. A

few of these may even today be spewing out great geysers of molten rock and gas, though we have yet to observe such an event occurring.

What would it be like to visit?

Besides being incredibly hot, Venus also has a dense atmosphere of very high pressure at its surface, about the same as the pressure at 1000 m (3300 ft) deep in the Earth's oceans. Such conditions cannot be endured by humans. The thick atmosphere would make it difficult to see objects very far away, and the carbon dioxide would, of course, be poisonous to breathe. The Sun would barely be noticeable from the surface, appearing as a yellow-orange smear seen through thick clouds. Venus has no moon, but even if it did, it would never be seen through the opaque clouds. Neither are stars visible at night. There are constantly raging thunderstorms on the planet, with very frequent lightning flashes. Most of the planet is fairly flat and covered by ancient volcanic lava flows, with some dust and gravel.

Earth – the oasis

Earth is the only planet in the Solar System that can support human life. It is a very pretty plan-

EARTH – Fact Box

Average distance from the Sun: 150 million km (93 million miles)
Diameter: 12,756 km (7926 miles)
Mass: 6×10^{24} kilograms
Time to orbit the Sun: 365 Earth days
Rotation time (a day): 23 hours 56 minutes 4 seconds
Number of moons: one

▶ *This photograph of a half-illuminated Earth was taken on Christmas Eve 1968 by Apollo 8, which was the first human flight to orbit the Moon. The day–night terminator is seen crossing Australia. India can be seen at upper left. The diversity of life on Earth is remarkable: more than 1.5 million species have so far been discovered and named.*

et, with swirling white clouds, blue oceans and mostly brown continents. No other planet in the Solar System has oceans of liquid water, plenty of oxygen in its atmosphere, and known life. The oceans cover more than two-thirds of the Earth's surface. The Earth has a thin crust of rock, which is slowly but constantly shifting. Violent proof of this plate movement is seen as earthquakes and volcanoes. Beneath the crust lies a 2700-km-thick (1670-mile) layer of denser rock called the mantle. At the center, the Earth's core is mostly made up of nickel and iron. The inner part of the core is solid and the outer part is molten liquid. The Earth's magnetic field, which causes compasses to point to the north, is generated in this metallic core.

The Earth's atmosphere today has more oxygen than when it was first formed over 4 billion years ago. Microscopic life in the Earth's early oceans converted gases into the air that we breathe today. More than three-quarters of Earth's atmosphere is now nitrogen, and most of the rest is oxygen.

Intelligent life on Earth also sets it apart from other planets. Thousands of artificial satellites orbit our planet. They help humans to communicate, monitor the weather, and even keep an eye on each other. An enormous variety of other living creatures also populates the Earth. In contrast, not even one living organism has so far been found on any other planet or moon in our Solar System.

What is it like to visit?

Earth is, of course, the ideal planet for humans. Gravity holds us down on its surface, oxygen in the atmosphere allows us to breathe, and the Sun keeps us warm during the day. Looking around, you can see attractive plants, birds, fish and mammals. The skies are constantly changing, with clouds forming and moving, rains falling, and perhaps the occasional lightning strike or hurricane. At night we can look through the planet's atmosphere and see stars and galaxies that are trillions of kilometers away. We can clearly see our own companion in space, the Moon, as well as other planets of the Solar System.

Earth's moon – our traveling companion

Except for Earth and Pluto, all the planets in our Solar System are considerably larger than their moons. Earth, however, is barely four times larger than its moon. In fact the Earth and Moon are sometimes called twin planets. This pairing has caused much debate as to how the Moon was formed. The theory that most scientists favor is that a body approximately the size of Mars collided with the Earth some 4.5 billion years ago, which is about the age of the oldest rocks collected from the Moon. The debris caused by this gigantic collision then gathered under gravity to form the Moon.

The Moon has no atmosphere and no oceans of water. In 1997 the Clementine spacecraft discovered ice inside craters on the Moon, which are likely remains of comets that crashed there billions of years ago. The surface of the Moon is pitted by numerous craters produced by a series of impacts about 4 billion years ago. Since the Moon has no atmosphere, rain or wind, these craters do not erode with time. The Moon is not completely dead, however, and Apollo astronauts have

◄ Unlike Venus, the Earth is still a very dynamic and active planet, as demonstrated by the powerful volcanic eruption seen here of Mount St. Helens, Washington, United States, in May 1980.

▼ This breathtaking view shows the full Earth rising over the Moon's north pole. The color image was taken by the Clementine spacecraft, which spent two months in 1994 orbiting and analysing the composition of the Moon.

MOON – Fact Box

Average distance from the Earth: 384,400 km (238,855 miles)
Diameter: 3475 km (2160 miles)
Mass: almost 1/100th of the mass of the Earth
Rotation time (a day): 27 Earth days, 7 hours and 45 minutes, with respect to the stars

To visit the Moon you need a spacesuit to carry air to breathe, and also to protect you from the Sun's intense radiation. You would not notice the heavy spacesuit, though, since the Moon has a weak gravity, which makes things feel six times lighter. The sky is black even during the day because there is no atmosphere to scatter sunlight. If you are lucky you might see the magnificent sight of Earth rising above the horizon and shining in the sky.

Mars – the red world

Mars, the red planet, has inspired numerous science fiction tales as a source of hostile alien beings and underground colonies. This planet is quite similar to Earth in that it has a thin atmosphere, polar ice caps, and dried-up river beds crisscrossing its surface. But Mars is much smaller and colder than the Earth. Another big difference is that there are no signs of civilization on Mars, past or present.

In August 1996 scientists announced the discovery of a meteorite found in Antarctica on Earth, which is believed to have been blasted off

▲ *The commander of Apollo 16, John W. Young, leaps from the surface of the Moon as he salutes the US flag. The lunar landing module is seen on the left, and a roving vehicle is parked beside it. The manned Apollo lunar landings gave scientists remarkable opportunities to explore an alien world.*

▼ *This spectacular global view of Mars was taken from the Viking Orbiter in February 1980. The perspective is from about 2500 km (1500 miles) above the surface. A huge rift valley called Valles Marineris extends from west to east for more than 2500 km (1500 miles). At its widest the canyon is 600 km (370 miles) across. Three ancient and extinct volcanoes can be seen on the left edge.*

recorded small quakes, which are the Moon's version of earthquakes.

From Earth, we always see the same side of the Moon. This effect is called synchronous rotation, meaning that the lunar rotational period coincides with its orbital period, and it is caused by the effects of gravity between the Moon and Earth. This gravitational attraction is also strong enough to pull the water on the Earth's oceans slightly toward the Moon, creating the tides on Earth.

What is it like to visit?

Between 1969 and 1972 six Apollo missions sent 12 astronauts to the Moon's surface. They gathered in total almost 400 kilograms of soil and rock samples and brought them back to Earth for detailed studies in laboratories. The astronauts found that the Moon's surface was strewn with loose rocks, ranging in size from pebbles to boulders as big as houses. They walked on dusty soil made up of loosely packed bits of rock and small glassy mineral deposits.

▲ *This panorama of the Mars Pathfinder landing site was taken in July 1997. Modest-sized hills (about 30 m/100 ft tall) are seen almost a kilometer away. The scene also includes ridges and a diversity of rocks.*

MARS – Fact Box

Average distance from the Sun: 228 million km (142 million miles)
Diameter: 6794 km (4221 miles)
Mass: 0.11 times the Earth's mass
Time to orbit the Sun: 687 Earth days (or 1.88 years)
Rotation time (a day): 24 Earth hours and 43 minutes
Number of moons: two

▶ *On 25 August 1998 Mars Orbiter cameras captured this wide-angle view of Olympus Mons, which is the tallest volcano on Mars. Its base measures 600 km (370 miles) across, and it rises to 26 km (16 miles) above the surrounding plains. Olympus Mons is the largest volcano known in the Solar System.*

from Mars after a large boulder slammed into the planet billions of years ago. The scientists reported signs of microscopic organisms (1000 times smaller than a human hair), which may have lived on Mars more than 3 billion years ago. At that time Mars was much warmer and wetter than it is now. These findings are controversial, however, and many people have proposed alternative explanations for the microscopic features, such as contamination on Earth.

Early in 2004 the Mars Express mission will place a lander called Beagle 2 on the surface of the planet. Beagle 2 will bore inside rocks and burrow into soil to search for evidence for primitive past and present life.

Mars is a small rocky planet. It looks red because iron in its surface soil has combined with oxygen to make rust: its surface is literally rusting away. Today Mars has only a very thin atmosphere, mostly of carbon dioxide, some of which condenses to make polar ice caps. Its surface is marked by some of the most awesome geological features. Valles Marineris, for example, is a gigantic equatorial rift valley that is deeper and wider than the Grand Canyon in the United States and longer than the distance from Los Angeles to New York. Mars is also home to Olympus Mons, the tallest volcanic mountain in the Solar System. It is 26 km (16 miles) high, which is almost three times the height of Mount Everest.

Mars has been studied in close-up by numerous spacecraft, including the Viking landers in 1976, which were the first to send back pictures from Mars' surface. In July 1997 NASA's Mars Pathfinder mission landed on the planet and released a robotic vehicle called Sojourner. Controlled remotely from Earth, it wandered around for weeks sending back new information about different rock types, soil composition and weather on Mars.

Mars has two tiny moons, barely 10 km (6 miles) across, called Phobos and Deimos. The gravity of these moons is so small that if an average human was to throw a ball hard upward from their surface, it would go into orbit. Phobos is slowly moving closer and closer to Mars, and it will shatter under the effects of gravity in another 100 million years or so.

What would it be like to visit?

A spacesuit would be needed on Mars, to give air to breathe and to protect from the Sun's harmful rays. You would also need the suit to keep warm. For long periods temperatures on Mars rarely climb above freezing and can drop to as low as −140°C (−220°F). Enormous sandstorms occasionally sweep the plains, blocking out sunlight and darkening the planet for days. The terrain is rather desolate, with a fine-grained soil littered with rocks of all sizes. Beyond the horizon, the daytime sky is a pale yellow-pink color, owing to the large amount of dust in the air. The larger of the two moons, Phobos, would look about half as big as the full Moon does on Earth. Over the next few years several spacecraft will be launched toward Mars, and this planet is likely to be the first one to be eventually visited by humans.

Jupiter – king of planets

Jupiter, the largest planet in the Solar System, is by far more massive than all the other planets put together. In many respects, including composition, it resembles a tiny star. Jupiter is a gas planet and does not have a solid surface like the Earth. From Earth we can only see the top layers

JUPITER – Fact Box

Average distance from the Sun: 778 million km (483 million miles)
Diameter: 142,984 km (88,846 miles)
Mass: 317 times the Earth's mass
Time to orbit the Sun: 11.9 Earth years
Rotation time (a day): 9 Earth hours and 51 minutes
Number of moons: at least 28

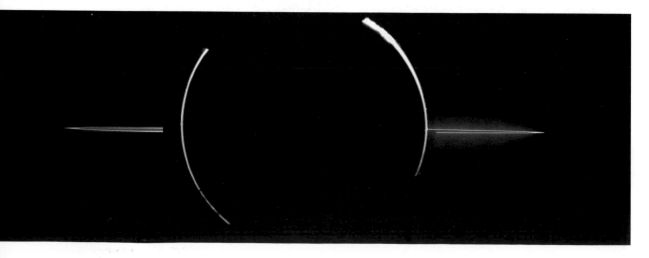

◀ *This eerie view of Jupiter's faint rings was obtained by NASA's Galileo spacecraft as it floated into the planet's shadow and peered back toward the Sun.*

of Jupiter's atmosphere. These layers form swirling patterns of clouds, with bands of different colors, carried in winds of several hundred kilometers per hour. Deeper down, and invisible from Earth, are likely to be vast, strange regions of liquid and metallic hydrogen, which are squeezed to incredible densities by the planet's gravity. Jupiter's atmosphere bristles with lightning and huge storm systems. The most famous of these is the Great Red Spot, which is a storm that has persisted for more than 300 years. Two bodies the size of the Earth could fit across this storm. The orange and brown colors seen in Jupiter's clouds are due to chemicals such as methane, sulfur and phosphorus.

◄ This image of Jupiter's dynamic moon Io was taken at a distance of 2 million km (over a million miles) away from the Galileo spacecraft. Io is covered with colored layers of sulfur compounds ejected 100 to 200 km (60 to 120 miles) high by active volcanoes. The surface has changed substantially over approximately 20 years.

◄ Jupiter casts a watchful eye toward its largest moon, Ganymede, in this stunning image taken in November 2000 from the Cassini spacecraft. Jupiter's Great Red Spot can be seen, together with smaller white ovals of rotating high-pressure regions in the giant gas planet's dynamic atmosphere. The smallest features visible are about 240 km (150 miles) across. Ganymede is about 50% larger than our Moon, and larger than the planet Mercury. The Cassini spacecraft used the gravity of Jupiter to get a pull toward Saturn, where it is due to arrive in 2004.

Jupiter has at least 28 moons, of which 11 small ones were only discovered in 2000. The Italian mathematician and philosopher Galileo Galilei discovered the largest ones, Io, Europa, Ganymede and Callisto, in 1610, using one of the first telescopes invented. Io is one of the most remarkable moons in the Solar System: it has active volcanoes on its surface, which frequently spew out yellow and red sulfur dust hundreds of kilometers above its surface. Ganymede is the largest moon in the Solar System – it is even larger than the planets Mercury and Pluto. Grooves and ridges crisscross its icy surface, which is also pitted with craters.

The two Voyager spacecraft sailed past Jupiter in 1979, not only providing spectacular close-up views, but also discovering three new moons and a faint ring system around the planet. The rings form three main bands, which are probably the debris of impacting bodies billions of years ago. They consist of tiny particles of very fine dusty material. In December 1995 the Galileo spacecraft dropped a probe into the atmosphere of Jupiter. It found conditions to be drier, hotter and windier than expected. However, the tiny probe only sampled a very small region of the giant planet's outer atmosphere. The orbiting Galileo spacecraft also discovered that Jupiter's satellite Europa has "warm ice" or liquid water below its rocky crust. Scientists and fiction writers have already started to speculate about the exotic organisms that might live in these oceans.

What would it be like to visit?

Jupiter's upper atmosphere of methane and ammonia is poisonous, and the massive planet's gravity is so strong that a human would find it very hard just to move. It is a gaseous planet, so there is no solid ground under the clouds to stand on. There are also roaring winds and very violent lightning. The temperatures can change rapidly from much hotter to much colder than any place on Earth. Jupiter rotates rapidly on its axis such that daytime lasts for only 5 hours – but you will not know this below the clouds, since these are too thick for sunlight to penetrate. No human would survive the crushing pressure of Jupiter's gravity.

SATURN – Fact Box

Average distance from the Sun: 1430 million km
(889 million miles)
Diameter: 120,536 km (74,898 miles)
Mass: 95 times the Earth's mass
Time to orbit the Sun: 29.5 Earth years
Rotation time (a day): 10 Earth hours and 48
minutes
Number of moons: at least 30

Saturn – jewel of the Solar System

Saturn, the sixth planet from the Sun, is a symbol of majesty and mystery. It is famous for its dazzling and complicated rings, which can be seen from Earth through a small telescope. Only Jupiter is larger than this giant gas planet. Saturn is mostly composed of hydrogen and helium, and like Jupiter it has no solid surface immediately below the clouds we see from Earth. As Saturn spins on its axis, winds force all the clouds into horizontal bands circling the planet.

Saturn's rings span about 270,000 km (168,000 miles) from edge to edge, which is more than two-thirds the distance from the Moon to Earth. The rings are very thin, however, perhaps only a few tens of meters thick. They are mostly made of bits of ice and rock, which range in size from pebbles to boulders. From 1980–81 the Voyager spacecraft revealed that

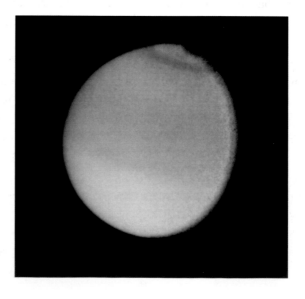

◄ *This view of Saturn's giant moon Titan was taken by Voyager 2 from a distance of 4.5 million km (almost 3 million miles). Titan has a thick, unbroken atmosphere rich in methane, nitrogen and hydrocarbons. Some of these molecules are the building blocks on which life on Earth was based.*

◀ This photograph of Saturn was taken in 1981 by Voyager 2. A shadow is cast by the rings on to Saturn itself, and the shadow of the giant planet is in turn seen on the rings. Saturn's rings consist of numerous icy chunks, which range from the size of pebbles to houses, all residing in thousands of narrow, closely spaced ringlets.

Saturn's ring system was in fact very complex, containing hundreds of narrow ringlets of varying brightness, each with numerous ice crystals. The rings most likely formed billions of years ago, either at the same time as Saturn itself or a little later, when some of its moons may have collided and split into the fragments that we see today. Saturn and its rings and moons resemble a miniature solar system, and studies of it can help us to understand more about the origin of our Solar System.

Saturn has more moons than any other planet – in the Solar System. Thirty of them have been discovered so far, the newest twelve being small satellites discovered since 2000. Each of Saturn's moons has unique characteristics, but Titan is possibly the most intriguing of all. Titan is larger than the planet Mercury and even has an atmosphere of nitrogen, which is similar to what we believe the Earth's atmosphere to have been like in the early stages of its development. Some scientists think that the surface of Titan may be covered by vast oceans of liquid methane. The joint National Aeronautics and Space Administration (NASA) and European Space Agency (ESA) spacecraft Cassini, launched in

1997, will arrive at the Saturn system in the year 2004. During its planned four-year mission around the planet, an instrument known as the Huygens probe will be dropped through Titan's dense atmosphere. If it survives its landing on Titan, the Huygens probe will send back readings from the surface, and it could provide astounding confirmation that vast methane oceans truly exist there.

Our understanding of Saturn is remarkable, but far from complete. There remain many unanswered questions about its huge atmosphere, complex rings and numerous moons.

What would it be like to visit?

Just like all the gas planets, Saturn would not be a welcoming place for humans. We cannot breathe its hydrogen atmosphere, and its ferocious winds would tear us apart. Even if we could travel through Saturn's clouds and survive the crushing pressure of its atmosphere, there would be no solid surface to stand on. Instead there would be an ocean of liquid hydrogen thousands of kilometers deep. From this deep interior you would be unable to see the Sun or Saturn's beautiful rings.

URANUS – Fact Box

*Average distance from the Sun: 2877 million km
(1788 million miles)*
Diameter: 51,118 km (31,763 miles)
Mass: 14.5 times the Earth's mass
Time to orbit the Sun: 84 Earth years
*Rotation time (a day): 17 Earth hours and 17
minutes*
Number of moons: at least 21

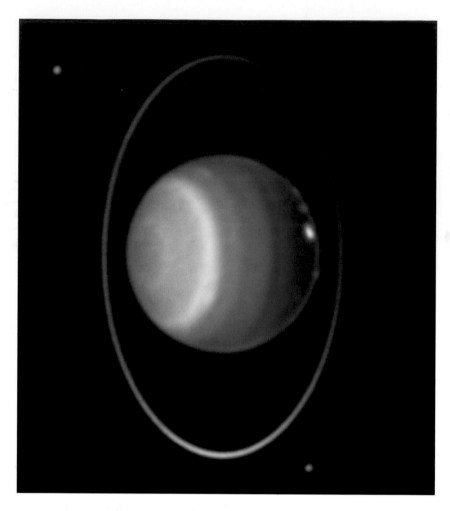

Uranus – rolling on its side

Uranus is also a giant gas planet. It has a cloudy atmosphere mostly made up of hydrogen and helium. Uranus is very unusual because, unlike any other planet in the Solar System, it spins sideways as it orbits the Sun. Astronomers believe that the planet was struck by a very large object billions of years ago, knocking it on its side. Uranus also spins on its axis in the opposite direction (known as retrograde) to most of the other planets.

Uranus appears little more than a tiny, pale green-blue disk through a large telescope on Earth. With the flyby of Voyager 2 in 1986, our knowledge of the planet increased dramatically. The atmosphere of Uranus has streaking bands and dark spots just like all the other giant gas planets. It is also circled by very thin rings, which are similar to those around Jupiter. These rings are composed of centimeter-sized particles and are too dark to be seen directly from Earth.

Uranus has 21 known moons. The largest is Titania, which measures 1610 km (1000 miles) in diameter. Another of its moons, Miranda, is one of the strangest moons in the Solar System. It is so heavily scarred, with craters, ridges, valleys and faults, that some astronomers think it may have been smashed apart at one time, then re-formed into a moon afterwards.

What would it be like to visit?

The winters on Uranus last for 42 Earth years on the north and south poles, and the Sun is not visible at all during these periods. You would not be able to breathe Uranus' atmosphere since it is poisonous, and you could not stand as there is no solid surface. The winds rage at speeds of 200–500 km/h (125–310 mph). Moving closer toward the planet the atmosphere gets thicker and thicker, until it changes from gas to liquid. Once again, a human would not be able to survive. Uranus' small central core lies beneath these impenetrable "oceans."

▲ *This false-color image of Uranus taken by the Hubble Space Telescope samples infrared light, which the human eye cannot detect. Several of the planet's satellites are seen, and four of its major rings. Clouds are also seen (pink-orange patches), circling the planet at more than 500 km/h (300 mph).*

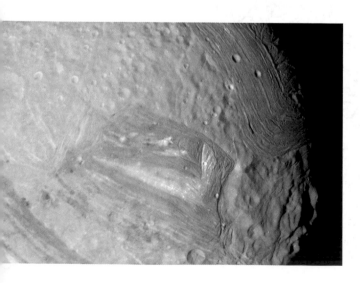

◄ *A Voyager 2 close-up picture of Uranus' satellite Miranda shows that its surface is heavily scarred by a complex network of valleys, ridges and canyons, which may be as much as 80 km (50 miles) wide.*

Neptune – blue world of the deep

Neptune is the fourth and last giant gas planet in our Solar System. It was discovered in 1846 after astronomers had calculated that the gravity of a heavy, unknown object was affecting the orbit of Uranus. Neptune's atmosphere is mostly made up of hydrogen. The uppermost regions that we can see from Earth contain methane gas, which gives the planet its stunning blue color. Streaks of white cloud, ranging in width from 50–200 km (30–120 miles), often cross the blue atmosphere. Several large oval-shaped storm systems have been spotted on this planet. In some ways Neptune looks rather like a blue-tinted Jupiter.

In 1989 the Voyager 2 spacecraft confirmed the presence of four dark, thin rings around Neptune. Although every giant planet in our Solar System has a ring system circling it, the properties of the rings themselves differ significantly from planet to planet. These differences lead to intriguing questions about exactly how the rings form and settle down, and the way in which the gravity of the planets' moons influences them.

Neptune has eight known moons, though many more smaller ones are likely to be discov-

NEPTUNE – Fact Box

Average distance from the Sun: 4498 million km
    (2795 million miles)
Diameter: 49,528 km (30,775 miles)
Mass: 16.7 times the Earth's mass
Time to orbit the Sun: 165 Earth years
Rotation time (a day): 16 Earth hours and 5
    minutes
Number of moons: at least eight

◀ *This striking image of Neptune was obtained by Voyager 2 in August 1989. The dark spot on the left is a giant storm system, measuring about 12,000 km (7500 miles) across. This spot had disappeared when the Hubble Space Telescope viewed Neptune in 1994, but a new storm was present in 1995. White wispy clouds made of methane ice crystals can also be seen. The presence of methane gives Neptune its beautiful bluish-green color.*

◀ *This Voyager 2 high-resolution image of Neptune was taken 2 hours before closest approach to the planet (a mere 4905 km/3050 miles from the planet's cloud tops). Vertical structure is seen in the bright cirrus clouds. Note the shadows cast by these clouds on to the lower layers of the atmosphere. The width of the cloud streaks ranges from 50 to 200 km (30 to 120 miles).*

ered in the future. The largest of these satellites is Triton, a fascinating object about one-third as massive as Earth's moon. Triton has methane and nitrogen on its surface, which reflects much of the sunlight that falls on it. This makes Triton a very cold place indeed, with temperatures at the surface being as low as −260°C (−436°F). Triton is also unusual in that it orbits around Neptune in the opposite direction to all the other satellites of this planet. Scientists think that Triton did not form at the same time as Neptune, but was instead captured later by the planet's strong gravity. Millions of years from now Triton's orbit will become too close to Neptune, and the effects of gravity will smash the moon apart. The debris of Triton will then scatter around Neptune, forming ring systems to rival those of Saturn.

What would it be like to visit?

Neptune's upper atmosphere of methane and ammonia is poisonous to breathe, and the weight of its atmosphere is unbearable. The Sun and Neptune's moons would soon become invisible through the clouds. Neptune's moons are no more hospitable to humans than is Neptune itself. Triton is colder than any other planet or moon in the Solar System. From Triton the Sun would look like a bright star, and Neptune would be an almost permanent giant blue feature in the sky.

PLUTO – Fact Box

Average distance from the Sun: 5915 million km
  (3675 million miles)
Diameter: 2274 km (1413 miles)
Mass: 0.002 times the Earth's mass
Time to orbit the Sun: 249 Earth years
Rotation time (a day): 6 Earth days and 9 hours
Number of moons: one

Pluto – baby of the family

Tiny Pluto is the last of the nine planets to be discovered in our Solar System. This planet is smaller than the Earth's moon and is the only planet never to have been visited by spacecraft from Earth, though scientists are considering a possible mission for launch within about the next 15 years. Pluto is very different from all the other planets. Its orbit around the Sun is highly elongated (or elliptical) and much more tilted than that of the other planets. In fact, some people wonder if Pluto should be considered a planet at all. For example, it had been thought to be an escaped moon from Neptune, though this idea is no longer favored. Some scientists now think that Pluto should be classified with a group of icy comet-like objects found at the far reaches of the Solar System, in an area known as the Edgeworth–Kuiper Belt.

◄ *These are the first maps of Pluto, made using direct images from the Hubble Space Telescope and enhanced using computer image processing. They show changes in large contrasting patches. These changes are probably due to different frosty regions of nitrogen and methane on the surface.*

A Hubble Space Telescope image shows the remote planet Pluto and its tiny moon Charon. The picture was taken in February 1994 when the planet was almost 4.5 billion km (almost 3 billion miles) from Earth (nearly 30 times the distance between the Earth and the Sun). Pluto is the only planet in our Solar System that has not been photographed by a passing spacecraft. The planet therefore remains somewhat mysterious.

Pluto is icy and rocky, with frozen surface layers of methane. The temperature at the surface is about −220°C (−365°F), but it will get even colder as Pluto moves farther from the Sun over the next few decades. Pluto has a moon called Charon, which is nearly half the size of the planet. These relative sizes make the pair similar to the Earth–Moon system.

What would it be like to visit?

Pluto is a very cold place indeed, since it is almost 40 times farther from the Sun than is the Earth. There is no oxygen to breathe, and the ground around is very icy and hard. Pluto is an extremely small planet and its gravitational pull is less than one-tenth that of the Earth, meaning that you would feel very light there. Its moon, Charon, always remains over the same spot on Pluto's surface, so you could only see it from one side of the planet.

Planet X?

Over the years, scientists have pondered about and searched for a tenth planet well beyond the orbits of Neptune and Pluto. None has ever been found, and certainly any extra planet at least as massive as Uranus and Neptune would have been discovered by now. The Pioneer and Voyager spacecraft have penetrated well beyond the orbit of Pluto, and there has not been any substantial drift in their paths that can be explained as being due to the gravity of a mysterious, massive tenth planet. Instead what lies beyond Pluto in our Solar System is a vast region containing billions of small icy and rocky objects.

Asteroids – interplanetary debris

There are almost 40,000 asteroids cataloged in the Solar System so far, and there are probably millions more awaiting discovery. Asteroids are small, rocky objects that orbit the Sun. Mostly they are part of the debris left over from the material that formed the planets billions of years ago. Asteroids can thus provide a wealth of information on the origin and evolution of the Solar System. Most asteroids are barely a kilometer or so across, though the largest ones, such as Ceres, can be almost 1000 km (620 miles) in diameter. Asteroids are the source of most of the meteorites we see on Earth. Meteorites are the tiny bits of rock that enter the Earth's atmosphere and survive all the way to the ground, rather than totally burning up. (Comets, however, are the sources of the meteor showers seen at certain times of the year.)

Most asteroids are found in a vast ring called the Asteroid Belt, which lies between the orbits of Mars and Jupiter. It is thought that the strong gravitational pull from Jupiter prevented these asteroids from clumping together to form a planet. Few asteroids have been studied at very close range. On its way to Jupiter, the Galileo space probe took close-up views of the asteroids Gaspra in 1991 and

▲ *The Galileo spacecraft visited the asteroids Gaspra (left) and Ida (right) on its way to study Jupiter. Irregularly shaped and pitted with craters, Gaspra is viewed from a distance of 1600 km (1000 miles). It has dimensions of only 19 by 12 by 11 km (12 by 7.5 by 7 miles). More than 600 craters, barely 100–500 m across (330–1650 ft), are present on this asteroid. Ida (right) is pictured from a range of about 10,500 km (6500 miles). It is about 50 km (30 miles) long and even more heavily cratered than Gaspra. Remarkably, this asteroid even has a tiny moon about 1.5 km (less than 1 mile) across, which is named Dactyl. This moon can be seen here to the right of the asteroid.*

Ida in 1993. Seen from a few thousand kilometers away, both appeared as lumpy, potato-shaped boulders, pitted with numerous small craters. In fact, they looked rather like Phobos and Deimos, the two moons of Mars, and some astronomers believe that these moons are actually asteroids trapped by Mars' gravity. Gaspra and Ida are about 20 km (12 miles) and 50 km (31 miles) long, respectively. A big surprise from Galileo pictures was that Ida even has its own tiny moon, Dactyl, just 1.5 km (less than a mile) across.

In February 2000, a spacecraft called NEAR Shoemaker became the first man-made satellite to orbit an asteroid. It carried out detailed imaging of the 32-km (20-mile) wide asteroid called 433 Eros. At times, the orbiting spacecraft got to within 40 km (25 miles) of the rocky surface. On 12 February 2001, the NEAR spacecraft actually landed on Eros and even survived long enough to transmit a short burst of gamma ray readings from the surface. Far-thinking entrepreneurs might one day visit asteroids in order to extract the rich metals and ores they contain.

Comets – visitors from beyond

A comet can be one of the most thrilling spectacles, with blazing tails streaking across the night sky. The comets derive their name from the Greek word *kome*, meaning "hair." They spend most of their lives as rather "dead," cold bodies in the outer reaches of the Solar System, well beyond the orbits of Neptune and Pluto. At certain intervals they visit the inner Solar System and come to life spectacularly as they near the Sun. Comets Hyakutake and Hale–Bopp were responsible for stunning displays seen in 1996 and 1997. More recently, Comet SOHO (C/1998 J1) was an exciting sight during mid-1998. It was discovered in May 1998 by the Sun-monitoring SOHO satellite. During late March 2002, Comet Ikeya–Zhang was a beautiful binoculars object, low in the western skies.

Comets can take anything between hundreds, thousands and even millions of years to make the trip round the Sun, in looping elongated orbits. The Earth sometimes passes through the long trails of tiny dust particles left by these visitors from the farthest regions of the Solar System. The result can be a spectacular meteor shower, with numerous brief streaks of light in the night sky, such as the Perseids in August and the Geminids in December.

A comet is a "dirty snowball" made of lumps of rock and frozen gas. Close to the Sun, its surface warms up and material boils off to form long, glowing tails of dust and gas. Comets are some of the oldest and best-preserved objects in the Solar System, dating back 4.5 billion years. Astronomers believe that billions of these objects reside in a vast shell called the Oort cloud, which lies beyond all the planets in our Solar System, at about a thousand billion kilometers from the Sun. It takes sunlight more than five weeks to travel this distance. Occasionally one of these lumps of icy rock is knocked out of the Oort cloud and sent in

◄ The majestic Comet Hale–Bopp is seen blazing across the sky during 1997. It features distinct tails of dust and gas, plus an extra one made of sodium. The tails of comets can stretch up to 100 million km (60 million miles) in length as they approach the Sun.

a long, looping orbit around the Sun. As we will see later in this chapter, some of these comets can even end up on a collision course with planets like Jupiter and Earth.

In 1986 Comet Halley was visited by several space probes from Earth. The Giotto craft of the European Space Agency (ESA) photographed the comet from close range. Giotto was expertly maneuvered, traveling at a speed of 70 km/s (44 miles/s), to come within 600 km (370 miles) of the center of Comet Halley. It discovered a 16-km-long (10-mile) by 8-km-wide (5-mile) object, with a surface darker than charcoal. Jets of gas were boiling off this surface. A spacecraft called Stardust was launched in February 1999 toward Comet Wild 2. It will aim to collect dust from the comet's tail in 2004 and return the samples to Earth for detailed study. The samples will teach us more about the composition of the Solar System when it formed billions of years ago.

Space weather
The forecast for tomorrow...

One of the great benefits of having a powerful instrument like the Hubble Space Telescope in operation for several years is that it can serve as an interplanetary weather satellite. For several years it has been observing the climate conditions on Mars and Venus. These studies are valuable since they can provide a better understanding of Earth's weather system. In the case of Mars' climate, predictions will also be critical prior to human exploration of the planet. When Mars has been closest to the Earth, the Hubble Telescope has distinguished details on the surface as small as 45 km

◄ On 13 March 1986 the Giotto spacecraft approached Comet Halley in a 600 km (370 mile) flyby. Its camera returned this unique image of the core of the comet. A dark, peanut-shaped body is seen, with two bright jets spewing out material. This nucleus of the comet is darker than coal and measures 16 by 8 km (10 by 5 miles).

(28 miles) across. This allows astronomers to track subtle shifts in cloud patterns and dust storms. One day we might be listening to weather forecasts like this:

Mars – A dry, cool and clear day, with low morning haze giving way to a mostly sunny afternoon. Some high clouds expected. A slight chance of severe dust storms driven by hurricane force winds. These storms may engulf the entire planet. The noon summertime temperature will be about 25°C (77°F), dropping to –150°C (–240°F) at night.

Venus – Another very hot and totally overcast day. Tremendous lightning storms are likely, with prolonged sulfuric acid rain. Temperatures will reach a sizzling 450°C (840°F). Smog levels remain dangerously high.

Jupiter – A new hurricane about the size of the Earth is raging in the northern hemisphere. Expect thunderstorms, lightning and rain. It's going to be phenomenally windy, with peak

speeds of gusts reaching almost 500 km/h (310 mph). The temperature at the cloud tops will be about –150°C (–240°F).

Neptune – Very powerful equatorial jet streams expected to blow at over 2000 km/h (1240 mph). Several hurricane storm systems will continue to rage. Some high-lying cirrus clouds of methane crystals are expected during the day.

Storms from the Sun

Our star, the Sun, provides an enormously powerful contribution to conditions in the space around the Earth and in our Solar System, an effect sometimes referred to as space weather. The Sun is a huge thermonuclear reactor, and its surface is a bubbling mixture of hot, electrified gas (plasma). Sometimes this activity increases and explosive bright spots called flares appear on the surface of the Sun. Single flares can release energy equivalent to 10 million volcanic eruptions on Earth. These awesome events can occur several times per year, and, together with other types of gigantic magnetic storms, can send huge bubbles of plasma heading toward the Earth. These bubbles travel at speeds of 400 km/s (250 miles/s), and the results for us on Earth can be spectacular and sometimes hazardous.

The normal "rain" of energetic particles in the plasma ejected from the Sun can enter the Earth's atmosphere and produce beautiful light shows called aurorae (the northern or southern lights). They are usually seen as a dazzling dance of green, blue, white and red light. Aurorae tell us that something electric is happening in the space around the Earth.

But there are also some disadvantages to this input of material from the stormy Sun. Giant eruptions can dump enough electricity into the Earth's atmosphere to more than twice meet the power generation capabilities of the whole of Europe. These electric and magnetic storms can wreak havoc on our communication satellites, electrical power and radio communications. For example, mass ejections from the Sun in March 1989 caused such a potent magnetic storm around the Earth that almost 1500 man-made satellites slowed

▼ These Hubble Space Telescope images of Mars were taken on 27 June 1997 (left) and 9 July 1997 (right). A large dust storm can be seen dispersing during the 12 days that separate these observations. The dust storm is the yellowish region at the 7 o'clock position in the two upper images. The storm is shown close-up in the lower pictures. Hubble observations show how rapidly weather conditions can change on Mars. The green cross marks the Pathfinder landing site.

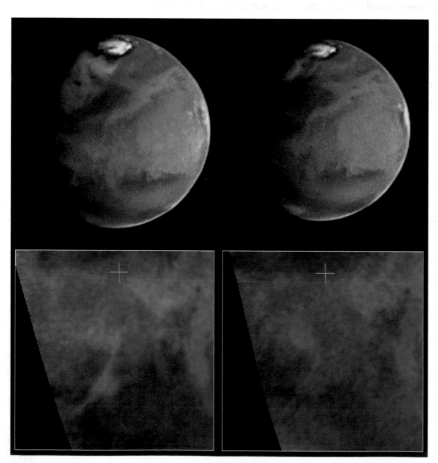

► These six images of the Sun were taken over a period of $3\frac{1}{2}$ hours by the Solar Maximum Mission. To enable a study of material being ejected, the central brightest regions of the Sun have been blocked out. Tremendous eruptions are seen swelling outward. These awesome events, known as coronal mass ejections, blast away charged particles at speeds of 100 km/s (60 miles/s) or faster. Such events may occur every few months or so, and the ejected material may reach the Earth a few days later. The resulting input of excess charged material (or plasma) into Earth's atmosphere can lead to spectacular displays of aurorae (northern lights).

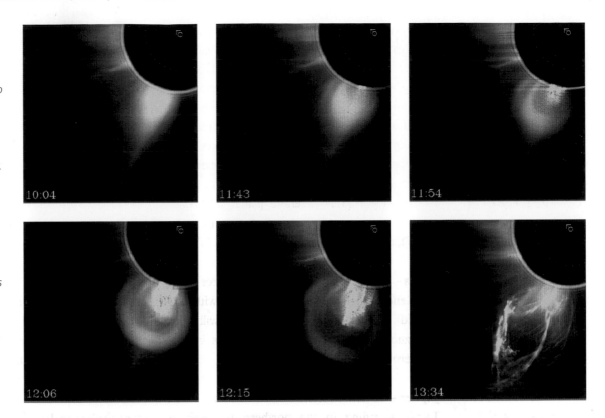

► This Hubble Space Telescope image taken in May 1994 shows Comet Shoemaker–Levy 9 already smashed into 21 pieces by the powerful gravitational force of Jupiter. The fragments spread over a distance of more than 1 million km (0.6 million miles), which is nearly three times the distance from the Earth to the Moon. Two months later these kilometer-sized chunks crashed into Jupiter.

down or dropped several kilometers in altitude. This radiation from the Sun can also cause defects in the solar panels used to power satellites and may even shut off computers on spacecraft. In 1994 two Canadian satellites were shut down when they were electrically shocked during magnetic storms caused by a turbulent Sun. In the most extreme cases, magnetic storms on the Sun can wipe out radio communications around the Earth's north and south poles for hours or even days.

We live in a satellite-dependent world, where each new rocket launch carries more sophisticated and sensitive electronic components than the previous one. But this also means that the satellites are becoming more vulnerable to the space environment. Scientists have, therefore, begun to monitor the Sun very closely and are continu-

ously watching for eruptions and storms. Their aim is to provide a reliable space-weather forecasting system one day.

Crashing comets
The collision between a comet and Jupiter

During seven hectic days in July 1994, over 20 pieces of a comet called Shoemaker–Levy 9 crashed into the giant gas planet Jupiter. The remarkable collision was watched by astronomers at hundreds of observatories around the world. The event was particularly special because it was the first time that a violent cosmic collision had been predicted, prepared for, and then actually observed.

Comet Shoemaker–Levy 9 was first discovered by the astronomers Carolyn and Gene Shoemaker

and David Levy in a photograph taken on 18 March 1993. The comet was by then already trapped in an orbit around Jupiter. Tidal forces caused by the massive gravity of Jupiter smashed the comet into 21 pieces. These fragments were spread out like a string of pearls, labelled A to W by scientists, excluding letters I and O. Pictures taken by the Hubble Space Telescope indicated that the largest pieces of the comet were about 2–3 km (1–2 miles) across, while the original comet was probably 8–10 km (5–6 miles) in size.

The first pieces struck Jupiter on 16 July 1994 and the last on 22 July 1994. The largest pieces were traveling at more than 200,000 km/h (125,000 mph), thus carrying a lot of kinetic (moving) energy. They dumped into Jupiter's upper atmosphere energy equivalent to more than 250,000 megatons of TNT, creating plumes that rose to 1000 km (600 miles) above Jupiter's cloud tops. Hot vaporized material from the pieces of the comet and Jupiter's atmosphere shot upward as if from a gigantic cannon.

But the fragment labelled G stole the show. It was about 3 km (2 miles) across, and it struck with an energy equivalent to 6 million megatons of TNT (that is more than 500 times the estimated total nuclear arsenal of the world). A fireball rose to a height of 3000 km (2000 miles), which is about a quarter of the Earth's diameter, and was seen through large telescopes around the world. Many of the impacts were ferocious and left temporary scars bigger than the size of the Earth on the upper atmosphere of Jupiter.

Overall, however, the giant planet itself was unaffected. Some scientists described the event as being like "lightly pressing 21 needles into a large apple" – after all, bits of ice and rock a few kilometers in size crashed into a gas planet more than 140,000 km (90,000 miles) across, and they only penetrated about 200 km (125 miles) into its upper atmosphere. From the thousands of images collected, and thanks to the natural probes provided by Shoemaker–Levy 9, astronomers are now beginning to learn more about the make-up of Jupiter's uppermost cloud layers. The event also taught us something about what comets can do when they collide at great speeds into a planet.

► Dark scars are seen in Jupiter's upper atmosphere after the impact of fragment G on 18 July 1994. This piece of the comet entered Jupiter at an angle of 45°. The Hubble picture shows a dark spot surrounded by a thin ring about 7500 km (4660 miles) across. The thick outermost crescent shape is almost 12,000 km (7500 miles) across (about the size of the Earth). The blemishes represent the "splash" from the impact, much like throwing stones into a lake. These scars in Jupiter's atmosphere persisted for several months after the impacts.

How safe is the Earth?

Civilization-threatening giant asteroids and comets sometimes play the leading role in blockbuster Hollywood disaster movies. The plot is not all pure fiction, however, and as we have just seen with Jupiter and Comet Shoemaker–Levy 9, violent collisions do occur in the Solar System. Spacecraft missions to Mercury, Mars and the satellites of Jupiter and Saturn returned pictures showing thousands of overlapping craters produced by impacts. The nature of these bodies resembles that of an artillery range. Pieces knocked off the Moon and Mars have even flown off and come to rest on Earth. Our planet itself has a surface relatively free of craters today, owing to the effects of erosion by rain and wind, together with geological upheavals like volcanoes. The Solar System is a much safer place now than it was billions of years ago when lots of free debris was flying about, yet to be captured around the planets.

Nevertheless, the Earth still glides in a slightly treacherous orbit around the Sun. It remains in a kind of cosmic shooting gallery, with a real possibility of impacts from comets and asteroids. For example, a group of asteroids known as the Apollo asteroids have orbits around the Sun that cross or come very near to the Earth's orbit. But, fortunately, the most common space debris to hit the Earth comprises tiny particles of space dust. These particles enter the atmosphere and are seen as streaking meteors, but they provide no threat to life. In fact, there has been no recorded case in the last 200 years of a human fatality caused by the impact of an object from space. The potential danger from cosmic impacts does, however, increase with the size of the projectile.

Significant impacts have occurred on the Earth in modern times. On 30 June 1908, for example, an incoming comet or asteroid exploded over the Tunguska region of Siberia. The blast flattened trees for 50 km (30 miles) around and blew enormous amounts of material into the Earth's atmosphere. The powerful event is thought to have been caused by an object only about 30 m (100 ft) in diameter, yet the force of its blast was equivalent to a 15-megaton nuclear bomb. Major crater-forming events on the Earth are rare, however.

Meteor Crater in Arizona, United States, is 50,000 years old, and the impacting body was about 50 m (165 ft) across. Scientists estimate that a 10-km-wide (6-mile) crater forms on Earth only about once in a million years.

If an impact is large enough, it can disturb the environment of our entire planet, with disastrous consequences. The most popularly quoted example is the impact suspected to have occurred about 65 million years ago, at the end of the Cretaceous period of geological history. It marks a period of mass extinction on the Earth, and the end of the age of the dinosaurs. Almost half the species on our planet became extinct at that time. The comet or asteroid that crashed into the Earth weighed more than a million tons and had a diameter of at least 10 km (6 miles). The actual impact site is thought to be the Yucatán region of Mexico. One hundred trillion tons of dust were lifted into the atmosphere, causing a drastic global climate change. Sunlight was blocked from reaching the surface, thus plunging our planet into a period of cold and darkness that lasted at least several months.

Although giant impacts are very rare, Earth-crossing asteroids only about 1 km in size could cause plenty of local damage. At present no asteroid or comet is known to be on a collision course with the Earth. However, of the suspected 2000 or so larger asteroids with orbits that cross that of the Earth, fewer than 200 have actually been discov-

▲ *An artist's impression of a giant asteroid striking into the oceans of the Earth. An impact of this kind would be powerful enough to disturb the environment of the entire Earth and cause ecological catastrophe.*

▲ *Barringer Meteor Crater in Arizona, United States, is the only relatively fresh large impact crater on Earth. About 50,000 years ago, a huge iron–nickel meteorite struck in Northern Arizona. The meteorite is estimated to have been 50 m (165 ft) across and to have had an explosive energy equivalent to a 20-megaton hydrogen bomb. The meteorite struck the ground with a speed of 40,000 km/h (25,000 mph) and created this crater, which measures over 1 km (just over half a mile) in diameter and is 200 m (660 ft) deep at its center.*

ered and tracked. Since there is a clear possibility that such impacts will occur in the future, they represent an important natural hazard for our long-term survival. The chance of a collision with a 1-km-sized object may only be as rare as once in a 100,000 years, but it does happen.

Postponing the apocalypse

It is, therefore, worth asking what could be done to save our planet from cosmic impacts. A small group of dedicated astronomers worldwide are systematically searching for objects in space that might threaten us. But to take a proper census of everything out there that is big enough and close enough is an enormous task. Major professional astronomical observatories are not suitable for carrying out such a long-term program, and there is not presently available the funding required to create new, more appropriate facilities. To some extent the sky scanning is left to the hard work of amateur astronomers. Meanwhile, we often remain in the dark, subject to unwelcome surprises. In 1994, for example, Asteroid 1994 YM1, a

tumbling, house-sized boulder, was discovered only one day before it flew past Earth at a distance of 100,000 km (60,000 miles), which in space terms is a near miss.

If an asteroid or comet should be discovered to pose a threat, what could be done to counter it? The most important factor will be how much warning time we have. If it was years away, we could perhaps land a probe on it and use rocket motors to alter the asteroid's orbit. If the warning time was much shorter, such as a matter of months, then the sole recourse might be a high-powered intercepting rocket armed with nuclear bombs powerful enough to blow it off course – though deflecting an object is more efficient than pulverizing it. Such a vision is rather unpleasant and very uncertain in outcome. An explosion may simply break up a 10-km (6-mile) asteroid into thousands of small fragments, perhaps increasing the likelihood of at least one of them impacting the Earth. The outcome may also differ from asteroid to asteroid. Some might be strongly held together, while others might be little more than lightly "glued" gravel heaps.

It is important to keep a sense of perspective over this issue. An exceedingly unlikely but nevertheless real threat is presented by cosmic impacts. In the long-term a careful, detailed astronomical survey of near-Earth space may be the key to surviving such a threat. Calls for thousands of nuclear-armed missiles on stand-by to deal with impacting bodies from space should certainly be treated with caution and reservation.

Planets around other stars

Throughout history humans have pondered over the possibility of other solar systems beyond our own. We live in an enormous Universe that has perhaps 100 billion galaxies that we can observe. Each galaxy contains hundreds or thousands of billions of stars. Many of these stars are very much like our Sun, in size, power and temperature. And the way our Sun and its family of planets is thought to have formed out of a collapsing giant cloud of gas and dust seems a fairly natural process. The odds are excellent, therefore, that there are other stars out there with families of planets.

Over the past few years, the whole issue of alien worlds around other stars has been dramatically lifted from science fiction to reality, with the discovery of planets circling nearby stars. These impressive discoveries have been made using advanced telescopes and clever techniques, and they have sent ripples of excitement throughout the astronomical community. But ignore any overzealous statements about life on these new-found bodies. They are in fact very unlikely to harbor life. The new planets found so far are strange worlds, most of which are unlike anything in our Solar System. Their discovery has even forced scientists to reconsider the way in which a star and planets form in the first place.

Finding new planets is very hard!

No one has actually seen the new planets; they were all discovered by indirect methods. It has been extremely difficult to find new planets because an observer cannot simply point a telescope at a distant star and see if it has planets going around it. Planets reflect very little of the light from their star, and they do not generate substantial energy themselves. In our Solar System, for example, the Sun is one billion times brighter than its

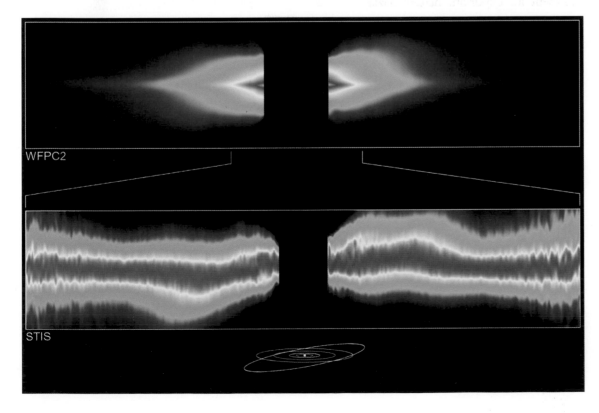

◄ Two Hubble Space Telescope images (false-color) show a disk seen edge-on around the star Beta Pictoris. Both views reveal warps in the dusty disk, which may be caused by the gravitational pull of planets still forming there. In both images the bright glare of the star itself is blocked out by the central black strip. The lower image shows the inner portions only of this disk, where the distortions are more clearly seen. For comparison, the orbits of the planets in our Solar System are also shown.

planets. So finding a planet directly around a star is like trying to find a firefly buzzing around a light-house beacon from a few kilometers away. We do not yet have the technology to do this.

Instead, the most common method used by astronomers is to look for wobbles in a star's motion. This wobble is due to the gravitational tug that a planet exerts on its parent star. As the planet revolves around the star, it pulls the star first one way and then the other. The more massive the planet and the closer it is to the star, the more noticeable its gravitational effect will be. The problem, of course, is that the planets are much smaller than the stars, and this wobble is tiny when viewed from a great distance. Someone gazing at our Sun from 30 light-years away would see it wobbling, because of the tug from its planets, by an amount that is equivalent to seeing a small coin

from about 10,000 km (6000 miles) away. However, with very precise and careful measurements, astronomers have now detected these wobbles of stars, which are recorded as shifts in the wavelengths of their spectral lines. They have come up with some startling new discoveries. In a few cases, it is also possible to detect a tiny decrease in light from a star as one of its planets passes across its face; this is known as a transit, and it is like a (very) partial eclipse of the star by the planet.

Over 70 planets have now been discovered orbiting other stars, and planet-hunting has really only just started. The ones discovered so far are essentially like the giant gas planets in our Solar System. They have between about half to ten times the mass of Jupiter. To everyone's surprise, these newly found planets exhibit two exceptional

▼ *An artist's impression of extrasolar planets. Currently we can only search for planets around other stars by indirect methods. Although we cannot directly image planets going around other stars, we can certainly speculate about what some of these alien worlds might look like.*

properties. First, unlike planets in our Solar System, which have mostly circular orbits around the Sun, several of the new planets move in oval orbits around their host stars. Second, many of the new planets orbit very close to their stars, some even closer than Mercury's orbit is to the Sun. Since the new planets cannot be seen, their composition is unknown.

The planet found orbiting the star 51 Pegasi is one of the oddest of the newly discovered planets. It has a mass almost half that of Jupiter, but it orbits at a distance of just 7 million km (4.5 million miles) from its star, which is less than one eighth of Mercury's distance from the Sun. It whizzes around 51 Pegasi so quickly that its "year" is just 4 days long. Being so close to the star, the planet has a likely temperature of about 1000°C (1830°F). Only slightly less strange is the planet discovered orbiting the star 47 Ursae Majoris in the constellation of Ursa Major (the Plow or Big Dipper). It is nearly two-and-a-half times the mass of Jupiter, but it orbits its star at a distance only slightly farther than that of Mars from the Sun. If placed in our Solar System, this new planet might appear as Jupiter's spectacular big brother. The nature of these hot and huge gas planets is not well established. It may be that they actually formed at large distances from their stars and slowly drifted inward.

Most of the extrasolar planets discovered so far have been the size of Jupiter or larger. More recently, however, astronomers have discovered two worlds smaller than Saturn. They are orbiting the stars HD 46375 and 79 Ceti. Another star, Upsilon Andromedae, harbors three Jupiter-like companions, making it the first system of planets found around another Sun-like star.

The current list of extrasolar planets represents only the tip of the iceberg. With time and new technologies it will be possible to discover many more planets around stars, giving us a better sense of the true variety of worlds out there. It may then be possible to compare our Solar System with other families of planets. We would learn if it is unique in some ways, typical, or one of many manifestations of nature's variety.

Today we can only barely detect the effects of giant planets around stars. There are plans being made to build powerful telescopes in space, which should be able to detect Earth-sized planets around normal (Sun-like) stars. The European Space Agency is, for example, developing the Darwin mission for launch around 2015. This giant observatory would have a base-line of at least 50 m (160 ft) and would be located somewhere between Mars and Jupiter. It will search for faint planets around 300 Sun-like stars located within 50 light-years of us, which is very close in astronomical terms. With these exciting new facilities we can anticipate far-reaching scientific discoveries, including whether there are other terrestrial-type planets out there with atmospheres and life forms. One thing is certain: we should expect to be surprised!

The life stories of stars

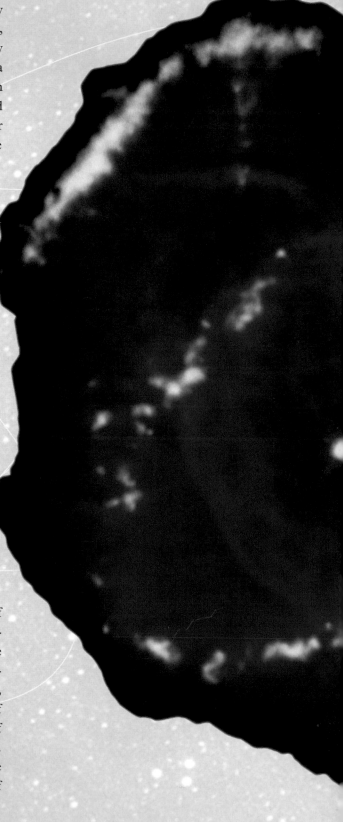

4

One of the most magnificent sights offered by nature is the starry sky seen on a clear dark night, far from city lights. On view is a bewildering array of stars scattered across the heavens, displaying a wide range in brightness and subtle variations in color. Almost 2000 stars may be seen by the naked eye, and a modest telescope increases this number into the millions. Astronomers estimate that there are in fact more than 100 billion stars in our Galaxy alone. Because we can see and study so many stars, it has become possible to group them according to their basic physical properties. They can be classified by their temperature, size, mass, how luminous they are, the chemicals they are made of, and so on.

By studying the starlight received in powerful and well-equipped telescopes, astronomers today know a great deal about the nature of the twinkling dots in the night sky. Some of these distant stars have bloated to hundreds of times the size of the Sun, while others glow with a radiance of more than a million Suns. Some stars are so remarkably crushed by gravity that they would easily fit inside a city the size of London. By surveying different groups of stars, and employing some clever detective work, scientists have come to understand that stars are not eternal and will not shine forever. Instead, stars have fascinating life stories, which unfold over billions of years. Majestic snapshots of the births, lives and deaths of stars have been captured by sensitive instruments like the Hubble Space Telescope, and this remarkable cosmic evolution is described in this chapter. Today, astronomers are able to extend our knowledge of stars from those in our Galaxy to populations of stars in other galaxies millions of light-years away. Many secrets of the nature and history of an entire galaxy can be unlocked by an understanding of the lives led by its billions of stars.

▶ The Ring Nebula, M57, in the constellation of Lyra, is a fine example of a planetary nebula. This nebula has been produced by the ejection of material from the outer layers of stars like the Sun.

What is a star?

A star is a gigantic ball of extremely hot gas held together by its own gravity. It is entirely gaseous. Unlike planets, a star emits light or radiation that is generated in its interior by nuclear reactions. (The light we receive from the planets and the Moon is not generated by these bodies themselves but is instead reflected sunlight.) The pressure of the hot gas in the star's interior counteracts the force of gravity, which is continually trying to collapse the star. As long as a star can generate its own power, it remains stable. We can only see the outer layers of a star; the interior where the energy is generated is hidden from our view by all the gas that surrounds it.

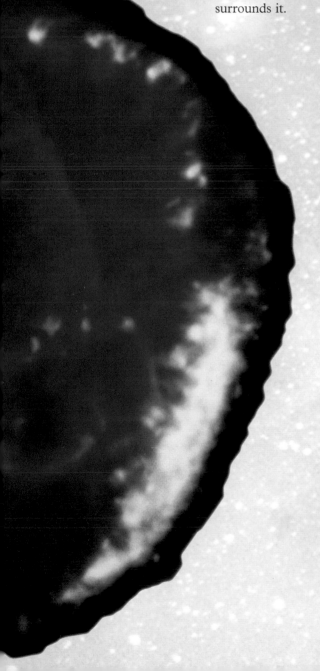

The Sun – an ordinary star

To discover more about the nature of stars, let's examine a very average one right in our own backyard – the Sun. We saw in Chapter 3 that the Sun completely dominates our Solar System. It weighs more than 700 times the mass of all the planets combined, and it has enough volume to hold more than a million Earths. The Sun is too hot to have a solid surface, and all the atoms and molecules contained within it are in the form of gas. Deeper in the Sun the gas is more compressed, and the pressure and temperature are greater. The temperature of the Sun rises from 5500°C (9900°F) at the uppermost layers to a remarkable 14 million °C (25 million °F) in the central core, where the pressure is almost a billion times that of the atmosphere on Earth. These are the requisite conditions for energy generation via nuclear fusion. The chemical make-up of the Sun is similar to that of most stars, with about 73% of its mass as hydrogen, 25% as helium, and the remainder as traces of heavier elements such as carbon, oxygen and nitrogen.

The fusion energy released in the core of the Sun is very slowly transported up to the outermost layers, from where it eventually radiates into space. From Earth, we can only see these upper

▲ *This beautiful star-filled view shows the dark night sky above the island of La Palma in the Canaries. The view is toward Sagittarius, and some well-known astronomical sights can be picked out, including the Lagoon Nebula (M8), the bright pink patch toward upper right. A luminous band of stars can be traced from the bottom right to the upper left; this is the Milky Way and it represents faint stars in the plane of our Galaxy.*

(or "surface") layers of the Sun, from which most of the energy is escaping as light. In the case of the Sun, we have a fantastic close-up view of a star, and we can see in great detail the dynamic and sometimes violent processes taking place in its outer layers.

A note of caution first: you should never look directly at the Sun as it can permanently damage your eyes. Only with the help of special filters and telescopes can astronomers study the principal surface layer of the Sun, which is called the photosphere, meaning sphere of light. (Remember that these surface layers are not solid but entirely gaseous.) Almost all of the Sun's visible light emanates from this very thin (400 km/250 miles) layer. Detailed images show that cells of hot gas and cooler gas are continually rising and sinking, respectively, creating ever-changing patterns. This process gives the photosphere its blotchy appearance. Sometimes darker patches called sunspots may be seen, which signal the effects of the Sun's strong magnetic field. Sunspots can be larger in size than the Earth. They are cooler by about 1500°C (2700°F) and hence darker by contrast than the

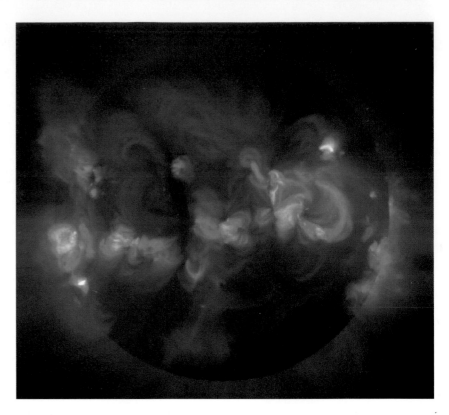

surrounding photosphere, which has a temperature of about 5500°C (9900°F). Sunspots can last several months, but most disappear in ten days or less.

▲ This dramatic X-ray image of a very active Sun was taken by the Yohkoh satellite in 1992. The brightest patches represent violent and energetic events where the gas has been heated to millions of degrees. The broader X-ray glow around the Sun is coming from its extended outermost layers.

◄ This picture of the Sun was taken from La Paz, Mexico, at the time of the total eclipse on 11 July 1991. The light from the brilliant "surface" (called the photosphere) is blocked by the Moon, making it possible to see the ghostly light of the tenuous outer atmosphere of the Sun, which is called the corona.

The layer immediately above the photosphere is called the chromosphere, and it has a temperature of about 10,000°C (18,000°F). Violent events can sometimes be seen here, the most awesome of which are solar flares, which are caused by the very sudden release of energy stored in highly twisted magnetic fields. The flares typically last only a few minutes, though the largest ones can persist for hours and emit enough energy to meet the power requirements of Europe at its current rate for the next 100,000 years. Huge, looping events known as prominences can sometimes be seen in the chromosphere. They can rise out to thousands of kilometers, and the material they eject majestically traces out the looping magnetic fields of the Sun. Variations in these energetic phenomena indicate that activity in the Sun fluctuates over a roughly 11-year cycle; for example, solar flares and large numbers of sunspots are more common during solar maxima, which occur approximately every 11 years.

One of the most remarkable views of the Sun is obtained during a total solar eclipse, when the Moon passes between the Earth and the Sun and by coincidence exactly blocks out the region of the photosphere. This gives us a rare chance to glimpse parts of the Sun not usually seen, since the glare from the photosphere normally dominates the view. For a couple of minutes or so during a total solar eclipse, we can see the ghostly white light of the outermost region of the Sun, the corona. The corona is at a temperature of 1 million °C (1.8 million °F). Scientists do not yet fully understand why this tenuous region is so hot, but it may be that the material in the corona is heated to such high temperatures by the dumping of energy stored in magnetic fields.

Finally, material (or plasma) from the corona expands outward in a flow of energetic atomic particles known as the solar wind. This electrified gas reaches the Earth at a speed of 400 km/s (250 miles/s). It is this solar wind that blows the tails of comets, as discussed in Chapter 3. The solar wind travels into the outer regions of the Solar System, carrying away material that has left the Sun. In this manner, the Sun loses about

The nuclear energy of the Sun and stars

Day in, day out, over the past 4.5 billion years the Sun has been generating tremendous amounts of energy. The brightness of the Sun (an ordinary star) is equivalent to 4 trillion trillion (4×10^{24}) 100-Watt light bulbs. It gets this prodigious energy from nuclear reactions in its core. The basic energy-generating process of the Sun and all stars is nuclear fusion, in which the nuclei of atoms merge to create a larger nucleus. All stars begin their lives generating energy from a fusion process known as hydrogen burning (this is nuclear burning, not combustive). The extreme temperature and pressure conditions in the core of a star create a nuclear powerhouse, in which four hydrogen nuclei are combined into one nucleus of helium in a chain of reactions. There is a simultaneous release of energy, together with subatomic particles called positrons and neutrinos. What is critical is that a tiny fraction of the mass of the hydrogen going into the nuclear reaction does not appear in the mass of the helium formed. This lost mass has been converted into the energy that makes the star shine.

The Sun is today converting 600 million tons of hydrogen into helium every second by this fusion process. It has enough hydrogen to continue generating energy at this rate for another 5 billion years. After that time, it will simply run out of hydrogen fuel and, as we will see later in this chapter, this will usher in a time of major adjustment.

▲ The Sun is a giant nuclear furnace. The energy that it emits comes from nuclear fusion reactions much like those that occur in a hydrogen bomb such as that shown here. A hydrogen bomb may produce an energy equivalent to more than 10 million tons of TNT. The Sun, however, releases almost 10 billion times this amount of energy in just one second.

► A panoramic Hubble Space Telescope image shows the stunning stellar nursery called the Orion Nebula. Shown here is a 2.5-light-year-wide view of the gas and dust that provides the raw material from which new stars and planets are made. The glowing turbulent material is illuminated by the intense light from a few extremely energetic young stars in the vicinity. The nebula is about 1500 light-years away from Earth, and is located in the middle of the "sword" region of the constellation of Orion.

The birth of stars

Stars are continually being made in the Universe today, and one new Sun-like star is born in our Galaxy almost every year. The birthplaces of all stars are enormous reservoirs of mainly hydrogen gas, plus dust composed of tiny grains the size of smoke particles or smaller. The giant clouds, called nebulae, are part of the material that lies between the stars (the interstellar medium) in a galaxy. These vast stellar nurseries may be hundreds of light-years across, and some can contain enough mass to make more than a thousand Suns.

A beautiful stellar nursery can be seen in the constellation of Orion, the hunter, about 1500 light-years away. Using a pair of binoculars, look toward the middle of the "sword" hanging from Orion's "belt;" here you will see a glowing patch of light – the Orion Nebula. Millions of years ago, dark clouds of gas and dust contracted here, under the force of gravity, to form a cosmic nursery containing luminous young stars. Often the gas and dust in a giant nebula like this can obscure direct views of the star-formation process. Nevertheless,

10 million tons of material each year. This is a fairly insignificant amount, however, compared to the total mass of the Sun. The most powerful stars in our Galaxy have very strong stellar winds indeed and can shed a billion times more mass than the Sun does every year.

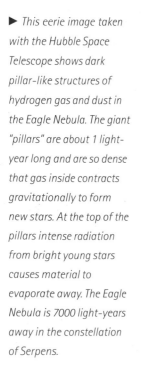

► This eerie image taken with the Hubble Space Telescope shows dark pillar-like structures of hydrogen gas and dust in the Eagle Nebula. The giant "pillars" are about 1 light-year long and are so dense that gas inside contracts gravitationally to form new stars. At the top of the pillars intense radiation from bright young stars causes material to evaporate away. The Eagle Nebula is 7000 light-years away in the constellation of Serpens.

astronomers have discovered dramatic evidence of star birth in recent years, using the sharp vision of the Hubble Space Telescope together with powerful telescopes that see in infrared light. Many images have been returned of pancake-like disks of gas and dust centered on the newborn stars.

All stars begin their life as a cloud of hydrogen gas and dust, and there are plenty of these in the Universe. The formation of a star begins with the collapse of the nebula under the force of gravity. Exactly how the collapse is triggered is not well understood, but one possibility is that it may be initiated by the blast wave from a distant exploding star known as a supernova. As the cloud collapses, it rotates faster and its center starts to heat up as it is compressed. After about 100,000 years we have a young contracting object called a protostar, which is still shrouded in an envelope of gas and dust. At this stage, a dense central condensation has formed, which will eventually become the new star. Powerful jets of material can sometimes be seen ejected from the protostar, traveling at velocities of hundreds of kilometers per second.

After about 1 to 10 million years most of the surrounding envelope of material has dissipated, and we get our first good look at the star. At this stage the stellar object is sometimes called a T Tauri-type star. Finally, after perhaps another 10 million years, the central core of the star gets hot and dense enough for nuclear fusion reactions to begin. The reactions convert hydrogen to helium, producing enormous amounts of energy, which is released as heat and light. The T Tauri-type star, which was a protostar, has now become a genuine star.

The moment of nuclear ignition marks the start of the most stable phase in a star's life. This phase is sometimes called the main-sequence phase, in which the star leaves its childhood behind and settles into a long middle age. The Sun is currently a stable main-sequence star, and it will remain so for another 5 billion years.

From youth to old age

Stars are not eternal and will not shine for ever. The simple reason for this is that stars are of a finite size. The Sun may be 100 times bigger

▶ *This is a Hubble Space Telescope image of a cluster of stars called NGC 1850, which lies 170,000 light-years away from us in the Large Magellanic Cloud. It shows a glorious collection of almost 10,000 stars blazing over a region 130 light-years across. This view toward the southern constellation of Doradus includes low-mass yellow stars, cool red giant and supergiant stars, and younger, hotter blue stars.*

than Earth, but it still has only a finite amount of nuclear fuel from which to generate energy. Eventually the Sun will exhaust its fuel, stop shining, and fade away, in the same way as a car coasts along the road and stops when its fuel tank is empty. Once a star has used up its supply of nuclear fuel, it must change and evolve. We will see shortly that these changes can be beautiful, dramatic or even catastrophic.

Fortunately for us on Earth, stars typically evolve over billions of years. Of course, as a typical person lives for less than a hundred years or so, astronomers cannot point a telescope at a single star and watch it go through all the stages of its life cycle. To understand how stars evolve, scientists must instead study an enormous number of stars, with different characteristics, spread throughout space today. Detailed study of their properties, plus considerable thought, leads scientists to discover that certain stars must evolve into other types of stars, and then into further different types, and so on.

As an analogy, imagine that an alien from another planet visited the Earth and had only one Earth day to study the life cycles of humans. Since one day is very short in comparison to the average life of a person, the alien will learn nothing by watching a single person during this period. Our visitor would do better to observe lots of different people, such as babies in a hospital, children at school, adults at work, and very old people. With data on these different age groups, the alien could then work out that humans must evolve from babies and toddlers to adults and then retired old people. It is in just the same way that astronomers play the detective game of stellar evolution.

So what's going to happen to the Sun?

The fate of a star in the late stages of its life depends critically on the amount of matter and mass it has at birth, when it first formed out of the contracting gas and dust cloud. Let us consider first the fate of relatively lightweight stars like the Sun, since there are in fact far more low-mass stars in our Galaxy than massive super-powerful ones.

Astronomers calculate that in about 5 billion years from now the Sun will have used up its central supply of hydrogen fuel. At that time it will have a core composed of helium, as a result of the nuclear fusion reactions described earlier in this chapter. Without the energy from nuclear reactions to maintain the outward pressure, there is nothing to counteract the force of gravity, so the Sun's outer layers will begin to collapse inward. The contraction will cause the central temperature to rise. Eventually enough heat will be generated to supply energy to expand the outer layers. These layers cool down once they get far enough from the central heat source. In effect, the Sun will become inwardly compact and hot, but outwardly bloated and cool. It will become what is known as a red giant star.

This cooler star, with a surface temperature of 3500°C (6300°F), now shines reddish instead of yellow-white. It will be nearly 2000 times brighter than the Sun today, and it will have expanded out to almost 100 times its present size. When the Sun is a red giant star it will engulf the planet Mercury. Although Earth might just avoid being swallowed up, we would not survive. The huge Sun would fill a third of the sky as seen from Earth, and our planet would have a temperature of more than 1000°C (1800°F). There are many red giant stars easily visible in our night sky, such as Arcturus in the constellation of Boötes, and Aldebaran in Taurus.

▲ *Hundreds of stars are seen toward this cluster, called Messier 50 (M50), including several bright red giants. M50 is about 3000 light-years from Earth and about 20 light-years across. Many stars in our galaxy reside in tightly packed globular clusters or open clusters of the kind shown here.*

THE LIFE STORIES OF STARS

Eventually the helium core of the red giant Sun will have contracted and heated under the action of gravity until it reaches a temperature of about 100 million °C (180 million °F). The incredible heat and phenomenal pressure in the core now provide the conditions for a new stage of nuclear fusion to begin. It becomes possible to combine helium (rather than hydrogen) in nuclear fusion reactions to make carbon. The start of helium fusion generates tremendous energy and makes our Sun stable against the force of gravity.

Within a couple of billion years the helium in the core will all have been converted into carbon, and once again gravity will take the upperhand. The core will contract but will not be massive enough to generate enough heat to start nuclear fusion of carbon. So this is truly the end of the nuclear fuel supply of the Sun. Gravity will crush the core into a very dense and compact star called a white dwarf, about which more will be explained later in this chapter. Meanwhile, the outer layers of the Sun will be puffed off into space as a cloud of material moving majestically out in all directions. This cloud is called a planetary nebula.

Planetary nebulae signal the death of stars like the Sun. The name was given by early astronomers since through their modest telescopes these clouds looked like planets or disks. We know today that a planetary nebula has nothing to do with a planet. The ghostly nebulae are instead the outer layers of a star, blown off by a burst of energy. About 1500 planetary nebulae have been located by astronomers, and it is estimated that on average one new planetary nebula comes into existence each year in our Galaxy. There are probably about 10,000 planetary nebulae currently hidden away behind gas and dust in the far reaches of our Galaxy.

Planetary nebulae display a variety of stunning and colorful structures: some are spherical, others

▼ *Shown here is an artist's impression of the fate of the Sun and Earth about 5 billion years from now. Having exhausted the supply of hydrogen in its core, the Sun expands enormously to become a red giant star. The intense heat of the now very nearby Sun boils off the Earth's oceans, burns away the atmosphere, and leaves a lifeless, largely molten surface.*

are elongated in one direction, and many are rather irregular in shape. A single planetary nebula would dwarf our entire Solar System. The shells of material are typically one quarter of a light-year across and are expanding into space at speeds of 20 km/s (12 miles/s). They eventually get so big and thinly spread out that they can no longer be seen from Earth. This phase in the life of the Sun will last only about 25,000 years, which is tiny compared to the billions of years of its previous history.

The compact white dwarf star left behind is the final form of our dying Sun. To start with, the surface of this bizarre object may be at least 100,000°C (180,000°F). With the passage of billions of years, it will cool from white hot to yellow, and then red, eventually becoming virtually undetectable as a completely cold and black, burnt-out cinder.

Massive stars – going out with a bang!

There are stars in our Galaxy that start their lives with masses as large as 50 or even 100 Suns. These massive stars shine with the brightness of perhaps 100,000 to a million Suns. The ultimate fate of these exceptional "heavyweight" stars is very different from and more dramatic than that of "lightweight" stars like the Sun.

Heavyweight stars are those born with a mass of anything between about 10 and 100 times the mass of the Sun. The constellation of Monoceros (the unicorn) boasts two of the most massive stars discovered. Together they are known as Plaskett's Star, and they revolve around each other. Each is a blue giant star estimated to be some 55 times the mass of the Sun. Although the vast majority of stars in our Galaxy are more like the Sun, the heavyweight stars nevertheless play a vital role in the functioning of the Galaxy. We will see that these stars provide the origin of life-giving chemical elements such as carbon, nitrogen and oxygen.

High-mass stars live in the "fast lane." The more massive a star, the more ravenous its fuel consumption, and the shorter its life. All the evolutionary changes happen much more quickly for high-mass stars than for solar-like stars. The larger mass and stronger gravity demand more pressure to keep the stars stable, and thus they have to generate more heat, which speeds up all the phases in their lives. We saw earlier that the Sun will spend a total of about 10 billion years converting hydrogen into helium in the most stable (adult) phase of its life. In contrast, a star born with a mass of 10 Suns will complete this phase in only

▶ *Many beautiful planetary nebulae have been photographed by the Hubble Space Telescope. The examples shown here illustrate the variety of shapes, structures and colors of such nebulae, which signal the death of low-mass stars like the Sun. The different appearances relate to the different conditions of matter in the ejected outer envelopes of stars. The nebulae are typically 1000 times the size of our Solar System, and each harbors at its center a highly compressed white dwarf star.*

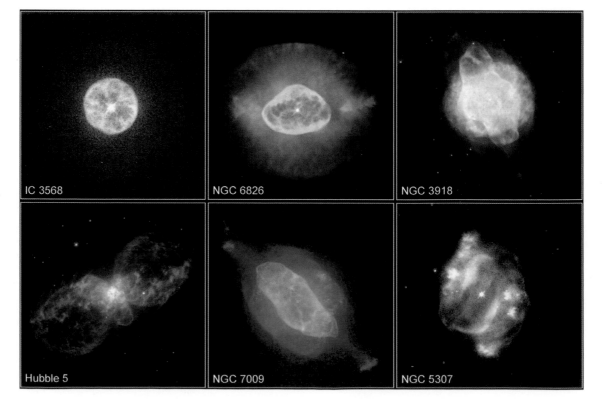

IC 3568

NGC 6826

NGC 3918

Hubble 5

NGC 7009

NGC 5307

◀ This is one of the most luminous stars known in our Galaxy. Called the Pistol Star, it releases up to 10 million times the power of the Sun and is big enough to fill the size of the Earth's orbit around the Sun. The Pistol Star gets its name from the shape of its surrounding nebula of ejected material, as seen in this Hubble Space Telescope image. The star itself is the bright object at the "trigger" of the pistol. This object is about 25,000 light-years away, toward the center of our Galaxy. The nebula is 4 light-years across, which is large enough almost to span the distance from the Sun to Proxima Centauri, the nearest star to our Solar System. Astronomers estimate that the Pistol Star was born about 3 million years ago with a mass almost 200 times that of the Sun. Using up its supply of nuclear fuel at a furious rate, the Pistol Star will likely die in a spectacular supernova explosion within the next 3 million years.

20 million years or so. So the tempo of a star's evolution is closely linked to its mass at birth.

The very important difference between heavyweight stars and lightweight stars is that once hydrogen has been converted into helium, and the helium into carbon, then (unlike the Sun) these stars can fuse heavier elements. Gravity will

◀ Eta Carinae is destined for destruction. The Hubble Space Telescope took this dramatic image showing some of the material the star has ejected over the course of its turbulent life. Historical records show that Eta Carinae underwent an unusual outburst about 150 years ago that briefly made it one of the brightest stars in the southern sky. With a birth mass of at least 100 times that of the Sun, this hot supergiant is an excellent candidate for a supernova explosion in the future.

We are made of stardust!

When the Universe came into existence some 15 billion years ago, the only elements present in great abundance were hydrogen and helium. The much heavier elements did not yet exist because they are the products of nuclear fusion inside stars. Almost every atom of every element, except hydrogen, helium and tiny amounts of other light elements, was made by nuclear reactions in stars. We have seen in this chapter how hydrogen in the core of a massive star produces helium, and helium produces carbon and oxygen, then silicon and so on. The remains of the nuclear burning of one element become the fuel for successive burning. As a star evolves, its nuclear-processed matter is ejected outward away from the dying star. Remarkably, this material eventually seeds a distant stellar nursery. It then becomes incorporated into the next generation of stars and solar systems.

The carbon in our brains, the calcium in our teeth, the oxygen we breathe, the silicon in the Earth's crust and the metals used in our industry were all produced billions of years ago inside stars. We are indeed made of stardust!

squeeze the matter in the core to reach temperatures of billions of degrees. The conditions will eventually be right to fuse carbon into oxygen, then eventually oxygen into silicon, and so on until the core is composed entirely of iron. It is not possible to fuse iron into a heavier element and release nuclear energy. Instead the fusion of iron actually requires the input of energy, exactly the opposite of what the star needs to maintain stability. The production of iron is thus truly the end of all possible nuclear fusion energy sources for any star.

At this stage the star will have an onion-like layered structure, with shells of differing composition produced by the previous stages of fusion. Somewhat similar to the Sun-like stars that swell

◄ A cluster of brilliant, massive stars called Hodge 301 is seen in the lower right corner of this Hubble Space Telescope image. The cluster resides in the Tarantula Nebula in our nearest neighbor galaxy, the Large Magellanic Cloud. Multiple generations of stars are seen in this nebula, including new stars being formed, red supergiants very close to the end of their evolution, and old stars that have exploded as supernovae and blasted material out into the surrounding regions.

out into giants, the high-mass stars become even larger as their outer layers are pushed away. Some become supergiants, perhaps several hundred times the diameter of the Sun. The stars Betelgeuse and Rigel in the constellation of Orion are both supergiants. To compare the size of the largest stars, imagine that the Sun is the size of a human eyeball – then the largest supergiants would be the size of a hot-air balloon.

Supernovae – the final act of destruction

Once a massive star has evolved to its final state, with a core of iron, it has no further source of internal energy. The size of the star, from the time that it first became stable, has been determined by an interplay between the inward force of gravity (trying to make the star smaller) and the outward pressure linked with energy production (trying to make the star larger). Once the star has an iron core, energy production comes to an end, and the star faces a dramatic and very violent collapse under gravity.

In less than one second the collapsing core undoes all the stabilizing effects of the previous 10 million years of nuclear fusion. The size of the core shrinks from a few thousand kilometers to about 20 km (12 miles) or less. The collapse is so fast that the outer layers of the star have no time to react and do not participate. The core temperature rises to over 100 billion °C (180 billion °F) as the iron atoms are crushed together. The collapse leads to an explosion called a supernova, which blows off the outer layers of the massive star. Like a fast-moving ball hitting a brick wall, the core becomes compressed, stops, then rebounds with vengeance. These events are remarkably rapid. Only about one second elapses from the start of the collapse to the rebound. The star explodes with far more violence than the expulsion of a planetary nebula that marks the end of a low-mass star like the Sun. The supernova produces a prodigious light show that can rival the power of an entire galaxy containing billions of normal stars. A typical supernova may shine brilliantly for several weeks, before fading gradually over many months. (Other types of supernovae arise not from the evolution of a single massive star but from the gravitational interaction of two stars in a close binary system. The final result is similar though – a gigantic explosion that destroys an entire star.)

Supernovae are fairly uncommon since the massive stars that spawn them are themselves rare. It is estimated that a supernova occurs in our Galaxy about once every 100 years. Many occur in distant parts of the Galaxy, however, and are not visible from Earth because they are obscured by the large regions of gas and dust present between the stars. Astronomers observe instead several supernovae each year in other galaxies, using telescopes and sensitive light detectors. The first nearby supernova to have been seen since the invention of the telescope in the early 17th century was observed on 27 February 1987; it was called Supernova 1987a (SN1987a). In the neighboring galaxy the Large Magellanic Cloud, which lies about 170,000 light-years away from us, a faint star brightened by over 200 times. Telescopes around the world and in space were directed toward the supernova to provide detailed information about the explosion.

The study of supernovae is not only of great interest to astronomers but also to physicists, who

▼ *Shown here are before and after pictures of the region of Supernova 1987a (SN1987a). The star that exploded in this very violent event is marked by an arrow. This brilliant light show in February 1987 gave astronomers their first chance to study the death of a relatively nearby massive star with modern scientific instruments.*

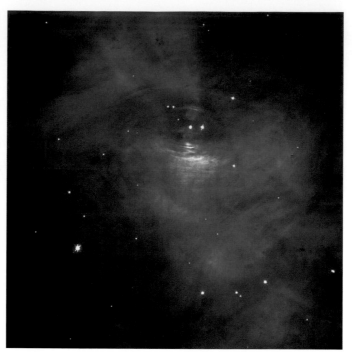

◀ *The detailed photo on the left is an image from the Palomar Telescope of the entire Crab Nebula, the remnant of a supernova first seen in AD 1054. The nebula is 7000 light-years away in the constellation of Taurus, and the ejected stellar material stretches 10 light-years across. A Hubble Space Telescope close-up image of the Crab Nebula's central region is shown on the right. The Crab Pulsar, which is a neutron star rotating at 30 times per second, is the left of the pair of stars near the center of the frame. (The other star in the pair is located much closer to us, and is not a companion in a binary system with the pulsar.) The wisps of material streaming away from the pulsar are traveling at half the speed of light.*

study the building of matter on the smallest scales. As spectacular as the supernova is in light output, most of the energy of the supernova is actually contained in ghostly (ultralight) particles called neutrinos. These particles are very difficult, but not impossible, to detect, and they tell the important story of the earliest phases in the collapse of the massive iron core. When SN1987a went off, scientists for the first time not only observed the light show but also detected 19 of the elusive neutrinos produced during the violent collapse. The burst of neutrinos preceded the first visual sighting of the supernova's light by three hours. This was exactly the time difference predicted, and the result was an exciting boost for scientific theories of supernovae.

Long after the initial supernova explosion, its aftermath can be seen as an expanding cloud of debris, known as a supernova remnant, produced by the remarkable destruction. One of the most famous supernova remnants is the Crab Nebula, the remains of the supernova of AD 1054. Recorded in Chinese literature, this supernova was said to be so bright that it could be seen in daylight and could be read by at night.

After the gigantic supernova explosion, all that remains of the star itself is a remarkably crushed core, typically containing two or three times the mass of the Sun. The material is at an even higher density than the white dwarf that will be the end state of the Sun. This special object is called a neutron star. Sometimes high-mass stars are so massive that no known force can halt the collapse of the core. Black holes are the result of the overpowering mass of such stars. We will see next that all the final states of stars are very strange objects indeed.

Stellar corpses

All stars have the same origin – a giant nebula of gas and dust – but after this stage nothing is common to all. We have seen that the life stories of stars depend crucially on how massive they are when they are born. A lightweight, Sun-like star has a long, stable middle age, then bloats out into a red giant, before eventually expelling its outer layers as a majestic planetary nebula. The star ends up as a white dwarf. Heavier stars consume nuclear fuel much faster. After an indulgent adulthood, they may turn into gigantic supergiant stars, before rapidly imploding and rebounding in phenomenal supernova events. The crushed cinders left behind in these cases may be neutron stars or, in the case of very high-mass stars, black holes. It is well worth taking a closer look at each of these extremely bizarre stellar corpses.

▼ *The brightest-appearing star in the sky, Sirius, is actually a binary star. Its companion (marked by the arrow) is a white dwarf star known as Sirius B. They are viewed here at X-ray wavelengths. Although Sirius A outshines it in visible light, Sirius B is far brighter in the X-ray waveband.*

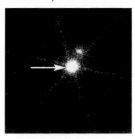

◄ *A Hubble Space Telescope image reveals several white dwarfs (circled) seen among a group of much brighter Sun-like stars. The detail shown here is part of a cluster of 100,000 stars called Messier 4 (M4), which lies about 7000 light-years away. Almost 40,000 of the stars in M4 are thought be be white dwarfs.*

White dwarfs

A star like our Sun will become a white dwarf when it has exhausted all its nuclear fuel. The white dwarf is initially very hot, perhaps 100,000°C (180,000°F) or more. It is typically half as massive as the Sun is today yet only slightly bigger than the Earth in size. White dwarfs are thus some of the densest objects known in space: if you could bring a thimbleful of white dwarf material to the Earth it would weight almost 50 tons. In this shrunken star the atoms are forced so close together that the orbits of the electrons are essentially destroyed and the atomic nuclei (protons and neutrons) could be described as being imbedded in a "sea" of electrons. This is a very special state of matter.

At least 60%, perhaps as many as 90%, of all stars end up as white dwarfs. Since they are so small and have such a faint output of light, astronomers have difficulty finding them. Sirius, the brightest-appearing star in the sky, is actually a pair of stars: the smaller compact star, Sirius B, is a white dwarf. White dwarfs have no way to keep themselves hot, so over the course of billions of years these small, subluminous, superdense, stable stars gradually cool down. Eventually they will cool completely and become very dark objects that do not radiate and cannot easily be detected: the white dwarfs become black dwarfs. In the process, the carbon in the core of the star crystallizes. As crystalline carbon is in fact diamond, we are left with "diamonds in the sky."

Neutron stars

Newborn stars with a birth mass of between about eight and 20 times the mass of the Sun will end their lives as neutron stars. The core left behind after a supernova explosion is highly compressed into a very weird state. With higher mass stars, the crushing action of gravity is strong enough to squeeze the protons and electrons together to form particles called neutrons. A neutron star is incredibly dense and extremely small. It is a sphere typically measuring only about 10 km (6 miles) across; it would easily fit inside London

▼ *This illustration shows the Earth, a white dwarf and a neutron star drawn on the same relative scale. In this case the white dwarf and neutron star have the same mass as the Sun.*

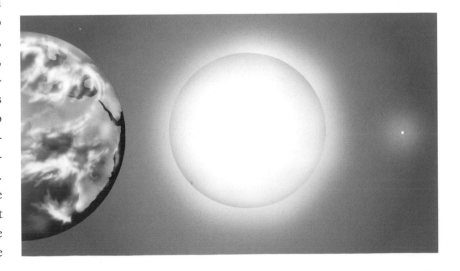

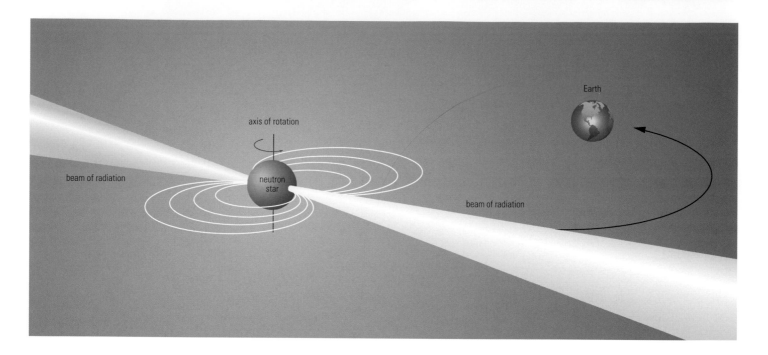

axis of rotation

beam of radiation

neutron star

beam of radiation

Earth

▲ *A pulsar is shown here in diagrammatic form. The magnetic poles of the neutron star are drawn perpendicular to the rotation axis in this case. As the star rotates, pulses of radio waves are emitted, and the beamed radiation sweeps the Earth. The mechanism is somewhat similar to a spinning lighthouse.*

or Berlin. However, its mass is greater than that of our Sun. The density of a neutron star is thus typically a billion times greater than that of a white dwarf. A thimbleful of its material would weigh as much as a large mountain on Earth.

Imagine an average human standing on the very thin crystalline crust of a neutron star. Because of the enormous gravity, this person would weigh the Earth-equivalent of a million tons and the relentless pull of the star's gravity would flatten him or her to about the thickness of a sheet of paper. It would, of course, be impossible to land on a neutron star in the first place. The strong pull of the star means there can be no soft landing: even a marshmallow dropped on to a neutron star from millions of kilometers away would crash into the surface with the energy equivalent to an explosion of a few million tons of TNT.

Neutron stars can also have very intense magnetic fields, perhaps 1000 billion times stronger than that of the Earth. Some neutron stars emit rapid radio pulses because of electrons being accelerated to very high speeds near the magnetic poles, which may not be aligned with the rotation axis of the star. Members of this exciting subclass of neutron stars are called pulsars.

A pulsar can be thought of as a sort of lighthouse, except the beacon of radiation is coming from a neutron star instead of a large light bulb.

The lighthouse-type radio pulses can be spaced by short intervals because these tiny stars may be spinning very rapidly indeed. A famous pulsar resides at the center of the Crab Nebula. The Crab Pulsar emits radio signals every 0.033 seconds, which means that the star is rotating once every 0.033 of a second, or 30 times every second. A pulsar called PSR 1937+214 rotates on its axis 642 times per second. Its surface therefore rotates at roughly one-tenth of the speed of light. In fact this pulsar is so stable and regular in its very rapid pulses that it can be used as an excellent standard clock for measuring time; it is better than an atomic clock.

Scientists have also discovered a breed of supermagnetic neutron stars, which they have named magnetars. Magnetars are extremely exotic and have magnetic fields perhaps 100,000 times stronger than those of most neutron stars. An object with such an enormously powerful magnetic field placed at a distance halfway between the Earth and the Moon could easily erase the magnetic strip of a credit card in someone's pocket on Earth.

Black holes – gravity wins

The strange behavior of tightly packed electrons and neutrons in white dwarfs and neutron stars, respectively, supports these two stellar end states from further collapse under gravity. The strangest stellar death of all occurs when the core left behind after the most massive stars undergo a supernova explosion is still roughly five or more times more massive than the Sun. (Remember these stars would have started life with anything between 20 and 100 Sun masses, but they have lost much of their mass through winds and explosive ejections.) Nothing in the Universe, not even compressed neutrons, is strong enough to hold up the remaining core against the force of gravity. The result of this collapse is the ominous black hole.

Out of all the weird and wonderful objects formed during the life stories of stars, a black hole is the most bizarre. Gravity scores the ultimate victory: it totally overwhelms all other known forces and crushes the core until it is infinitely small. This is a black hole. It is a region of space with so much concentrated matter that there is no way for a nearby object to escape its gravitational pull. To really understand the nature of a black hole, scientists need to use a theory that can describe the action of gravity under extreme conditions. Our best theory of gravity was put forward by Albert Einstein in 1916 and is called the theory of general relativity. Over the past few decades this theory has become essential in understanding many phenomena in the Universe.

According to general relativity, space itself becomes severely distorted near a black hole. This distortion is so extreme that even light cannot escape from the black hole. As an analogy, imagine a snooker or pool table made of a thin rubber sheet. Then place a heavy rock in the middle. The flat rubber sheet will sag, of course, and the warp or distortion in it will be greatest near the rock. Also, the heavier the rock, the larger the curvature of the rubber sheet. Now try playing snooker or pool! You will find that the balls passing near the rock get strongly deflected by the curved sheet. Balls that get too near the rock may drop down into the central sag and will never come out again. In the same way the space (rubber sheet) around a black hole (rock) is the most severely curved of all. The distortion is so extreme that not even light can escape from it.

Black holes are not, however, giant cosmic vacuum cleaners, steadily sucking up the entire Universe. Only objects that get very close to a black hole actually become trapped. The black hole may then draw in a whirlpool of matter, swirling around it, falling ever closer to the point of no return. The point where matter is trapped by the black hole and vanishes from sight is known as the event horizon. A black hole with a mass equal to that of the Sun would have an event horizon of only 3 km (less than 2 miles). Well beyond this distance, the black hole would exert a normal gravitational influence. So, if we were suddenly to replace the Sun by a black hole of the same mass, the Earth and the other planets would not get sucked in – they would keep on orbiting in exactly the same paths they follow right now.

Falling into a black hole

What would happen if you were able to stray too close to the event horizon of a black hole? Falling in feet first, there would be a stronger gravitational force on your feet than on your head, since Isaac Newton's theory tells us that gravity increases the closer you are to an object. The tidal forces would painfully stretch you apart. Eventually the forces would destroy every molecule inside you. All the material would be torn apart, heated to very high temperatures, and crushed to a single point. What would you see as you were falling in? You would not see anything particularly interesting, though images of faraway objects would be distorted in strange ways because the black hole's gravity bends light. Of course, no one outside the black hole would be able to see you, since light from you cannot escape from the black hole.

Detecting black holes

Since light cannot escape from inside a black hole, we cannot point a telescope and see one. The existence of black holes can only be confirmed by indirect methods. One of the principal ways of detecting black holes has been through observations of motions when a black hole and a normal star orbit one another under gravity.

In a system where a black hole and a normal star rotate around their mutual center of mass,

▼ *In this artist's impression of the binary system Cygnus X-1 material is streaming away from the normal (larger) star into an orbit around a black hole. As the gas spirals in toward the black hole (the other star in this binary system), friction heats it to millions of degrees and large amounts of X-rays are emitted. These X-rays can then be detected with telescopes in orbit around the Earth.*

astronomers can study the motion of the visible normal star in its orbit, and from this can deduce properties such as the mass of the dark companion. Sometimes this pair of objects may be so close together that material is pulled on to the black hole through a swirling disk. The friction of this material can make the disk so hot that it radiates tremendous amounts of X-rays, which can be detected using telescopes in space. The combination of the mass estimate of the invisible object together with analysis of the X-ray radiation can provide strong evidence that the object in question is indeed a black hole.

One of the most likely candidates known is a binary system called Cygnus X-1, a strong X-ray source in the constellation of Cygnus. It is thought to harbor a black hole with a mass about seven times that of the Sun.

Binary stars – living together

So far in this chapter we have dealt with stars as single, isolated bodies, leading their lives without external interference. Most stars in our Galaxy, however, are not solitary. At least half, perhaps two-thirds, of all stars in the Galaxy are in fact part of a co-orbiting system of two or more stars. A pair of co-orbiting stars is called a binary star. The study of binary stars is very important in astronomy, as these objects often provide vital information on the masses, sizes and outer layers of stars. In a binary star system, the two components may be separated by a fraction of a light-year, or they may be almost touching. (Recall that the nearest star to the Sun is over 4 light-years away.) A fascinating aspect of binary stars is that very different types of stars can be paired, including high-mass and low-mass, giant and dwarf, red and blue stars.

Astronomers believe that binary or multiple stars that are bound by gravity and orbit each other can be made in at least three ways.

First, it is possible that a large, single, newborn star (a protostar) is spinning so fast that it simply splits in two. The result would be a pair of stars, almost in contact with each other. Second, a massive single star may capture, by chance, another star that approaches close to it. This may happen, for example, where there is already a tight cluster of thousands of stars, with lots of scope for gravitational interaction. Many of the more widely spaced binary stars may form in this manner. Finally, in some cases the original collapsing nebula or stellar nursery may fragment to produce several moderately sized clumps, a few of which form stars close enough to become bound to each other.

The subsequent life stories of some binary stars can be far more complicated than those of single stars. For instance, a higher-mass star in a close binary system will evolve much faster than its low-mass companion. During its (swollen) red giant or supergiant phase the massive star will be virtually in contact with the lightweight star, and it can pull away swirling streams of additional material.

There are several different types of binary star in the stellar zoo of our Galaxy. One of the most fundamental types is called a visual binary. These stars are clearly gravitationally associated with each other, and they can be observed through a telescope to orbit each other around a common center. A well-studied visual binary is Castor, located in the constellation of Gemini. In order to see a visual binary, the separation of the two stars must be wide, and so the orbital periods are usually rather long, perhaps even as much as hundreds of years.

Another important category of binary system is that in which the components cannot be resolved visually because they are too close to each other. Such systems may be shown to be binaries by studying the nature of the combined light from both stars, which is passed through a spectrograph on a telescope. The orbital motion of the stars is revealed by periodic Doppler shifts

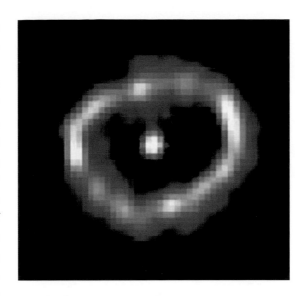

◀ In this Hubble Space Telescope image, a shell of material can be seen surrounding Nova Cygni 1992, which erupted on 19 February 1992. The nova is the result of a thermonuclear explosion that occurred on the surface of a white dwarf in this binary star system. The ring is the edge of a bubble of hot gas blasted into space. Nova Cygni is more than 10,000 light-years away, and it lies in the constellation of Cygnus.

of spectral lines in their composite spectrum. (The principle of the Doppler shift will be explained further in Chapter 6.) These binary systems are called spectroscopic binaries, and about 1000 of them have been detected. Since these stars are so much closer to each other than those in visual binaries, their orbital periods are much smaller. Algol, in the constellation of Perseus, is a spectroscopic binary, and the stars orbit every 2 days, 20 hours and 49 minutes.

Some binary stars get very close to each other to create bizarre effects. Earlier in this chapter, we mentioned the special class of stars called X-ray binaries. They are so-called because they produce a large amount of X-ray radiation. They comprise a normal star and a highly compressed one, such as a white dwarf, neutron star or black hole. The X-rays are produced because the stars get close enough together for material to be stripped off the normal star by the gravity of the collapsed denser star. An area known as an accretion disk forms around the collapsed star. This disk contains in-falling material heated to temperatures of over 1 million °C (1.8 million °F), and it radiates in the X-rays as well as other parts of the electromagnetic spectrum, such as the ultraviolet.

Finally, there exist exploding binaries called novae. An exploding binary is a pair in which the smaller star is a highly evolved and compressed white dwarf, and the larger star is a red giant. The red giant dumps hydrogen not used in nuclear reactions on to the surface of the white dwarf. The intense gravity of the white dwarf compresses and heats the hydrogen. Eventually, the hydrogen may ignite in nuclear reactions that suddenly blow this excess gas outward and away from the white dwarf. This type of explosion is known as a nova, and in some binaries it can occur in cycles perhaps thousands of years apart. (An example is shown on page 79.)

As many as 30 to 50 novae may explode in our Galaxy each year. If the red giant dumps too much gas on to the white dwarf too rapidly, the white dwarf may collapse to form a neutron star. In doing so it would blow off a fraction of the excess material in a phenomenal supernova explosion. The supernovae produced in close binary stars are called Type I supernovae; those created by the deaths of isolated massive stars are called Type II supernovae.

So what else can we learn from binary stars? Probably the most important feature of a binary star system is that it allows astronomers to determine the masses of the stars. This is particularly important because the mass of a single isolated star is very difficult to determine reliably. Furthermore, we have seen in this chapter that knowledge of a star's mass is crucial for understanding the way in which the star will evolve and what its end state will be.

Masses are determined from accurate measurements of the orbit of a binary star, including its size and period, combined with the laws of gravity and orbital motion. Sometimes the two stars in a binary system are aligned in such a manner that, when viewed from Earth, one eclipses (or exactly passes in front of) the other during its orbit. The duration of the eclipse together with data on the speeds at which the stars are traveling in their orbits can provide accurate measures of the sizes of the stars. By carefully watching one star sweep in front of the other, these eclipsing binary systems also enable different regions of the stars to be studied, such as the edges (or limbs) and the central regions. Such studies enable astronomers to improve their understanding of the structure of the upper atmospheric layers of stars.

Brown dwarfs – failed stars

When a giant nebula of gas and dust collapses under gravity to form new stars, the process can sometimes lead to the formation of protostars that do not have enough mass to become "proper" stars (without sufficient mass they cannot generate energy from nuclear fusion). These objects are called brown dwarfs and may be regarded as failed stars. If a star is born with a mass of slightly less than one tenth of the mass of the Sun, its cen-

The Top 10 of Stars

the 10 nearest to the Sun		the 10 visually brightest		the 10 largest	
name	distance (in light-years)	name	apparent visual magnitude	name	size (× Sun)
Proxima Centauri	4.2	Sirius A	−1.46	Epsilon Aurigae	2700
Alpha Centauri A	4.4	Canopus	−0.72	VV Cephei	1260
Alpha Centauri B	4.4	Arcturus	−0.05	Mu Cephei	1200
Barnard's Star	6.0	Alpha Centauri A	−0.01	Alpha Herculis	720
Wolf 359	7.8	Vega	+0.03	Mira	470
Lalande 21185	8.3	Capella	+0.08	Betelgeuse	400
Sirius A	8.6	Rigel	+0.12	Antares	360
Sirius B	8.6	Procyon	+0.40	Aldebaran	36
Luyten 726A	8.7	Achernar	+0.45	Arcturus	25
Luyten 726B	8.7	Betelgeuse	+0.50	Capella	20

▲ *A brown dwarf (the small white dot here) is seen orbiting the star Gliese 229 (the larger object in this Hubble image), with a separation similar to that between the Sun and Pluto. The diagonal spike is an artifact caused by the telescope optics. Gliese 229 is 18 light-years from Earth in the constellation of Lepus.*

tral temperature will never be hot enough to commence the fusion of hydrogen. Brown dwarfs are in some ways larger versions of the giant gas planet Jupiter. Typically, brown dwarfs have a mass range between 10 and 80 times the mass of Jupiter. They represent the class of objects that are too massive to be planets but too light to be stars. Brown dwarfs are thought to have temperatures in their central core of about 3 million °C (5.5 million °F), and their outermost layers would be at about 700°C (1300°F).

Brown dwarfs are very important objects in astronomy because they can easily be confused with the genuine extrasolar planets of the kind discussed in the previous chapter. It is important to learn enough about brown dwarfs so that they can be distinguished from planets. The usual notion is that brown dwarfs are objects that form in the same basic way as stars, only with insufficient mass to begin hydrogen burning. Planets, however, are formed from the material left over after the formation of stars. At the moment it is still rather difficult to confidently determine whether a distant object is a large planet or a small brown dwarf.

Scientists are also very keen to discover how many brown dwarfs there are in space. We will see in the following chapters that one of the most important issues facing astronomers today is the problem of the so-called missing mass. Detailed observations of galaxies have shown that we are only detecting the luminous radiation from about 10% of the mass of the Universe. So despite all our observations of planets, stars and galaxies, 90% of the mass of the Universe cannot presently be detected as electromagnetic radiation. So where is it all? One theory is that a fraction of the missing mass may be in the form of brown dwarfs. Since brown dwarfs do not have any nuclear energy sources, they do not shine brightly to begin, and get even fainter as they age. They can remain hidden in space and could, therefore, make a contribution to the missing mass problem of the Universe.

The problem in carrying out an accurate census of brown dwarfs in our Galaxy, of course, is how to find them in the first place. The luminos-

ity of a typical brown dwarf is only around a hundred-thousandth of that of the Sun, which makes it very hard to detect. The first one to be imaged directly, in 1994, was called Gliese 229B, and it was barely visible through the Palomar 60-inch telescope in California, United States. The object was subsequently observed by the Hubble Space Telescope, and it is thought to be about 30 to 55 times the mass of Jupiter.

The best places to look for brown dwarfs are in nearby star-forming regions, or places where there are many young stars, such as in stellar clusters like the Pleiades and the Hyades. One of the best hopes for directly locating brown dwarfs is to use special telescopes in space that can detect infrared light. These telescopes can detect the heat from brown dwarfs even though they are too cool to radiate in visible light. Alternatively, astronomers could use indirect evidence. They could, for example, search for the effects a brown dwarf would have if it drifted between us and a distant background star. If the brown dwarf is part of a binary system, then it may also be possible to monitor wobbles in the companion star as it is tugged by the gravity of the brown dwarf.

A cosmic recycling plant

This chapter has covered the remarkable story of star formation, evolution and death. This story also tells of a very important cycle that replenishes galaxies with nuclear-processed material. At the start of the cycle, stars form out of giant clouds of collapsing gas and dust. Their central regions or cores gradually get hot enough to create heavier elements from lighter ones by the process of nuclear fusion. During this process the star is evolving, and it changes its size, mass and power output. The most massive stars evolve the most quickly. All stars have methods of spewing some of this processed material back into space. So, through stellar winds, ejection of planetary nebulae or supernovae explosions, elements heavier than hydrogen and helium are seeded back into the space between stars. The material is put into the interstellar medium and into giant gas clouds. It is out of these giant

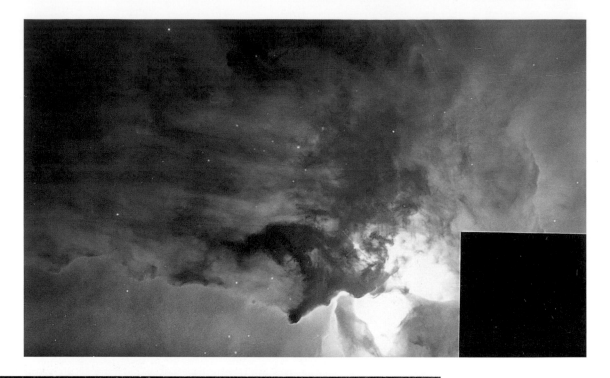

◄▲► *Four Hubble Space Telescope images are arranged to show the life cycle of stars. Above, giant clouds of gas and dust in the Lagoon Nebula are the site of new star formation. Top right, the stars in this dazzling array toward Sagittarius are blazing in glory and evolving along different life paths. Right, massive hot stars synthesize heavy elements by nuclear fusion and drive the processed material back into space in violent stellar winds. Left, expanding shells of material in the Veil supernova remnant contain heavy elements made in stars. These elements may be seeded into giant nebulae to become in turn part of the raw materials for the formation of the next generation of stars.*

clouds that the new – next generation – stars and planets will form, thus completing the cycle.

Our Sun is the product of many such cycles. When, almost 5 billion years ago, our star finally formed, the surrounding cloud of material not only contained large amounts of hydrogen and helium, but also certain amounts of metal-rich elements, such as carbon, oxygen, iron and so on. These were ejected into our cloud by previous generations of ancient stars and supernovae explosions. Without the synthesis of heavy elements in the hearts of other stars, planets and life on Earth would not exist. Our lives and those of stars are inextricably linked.

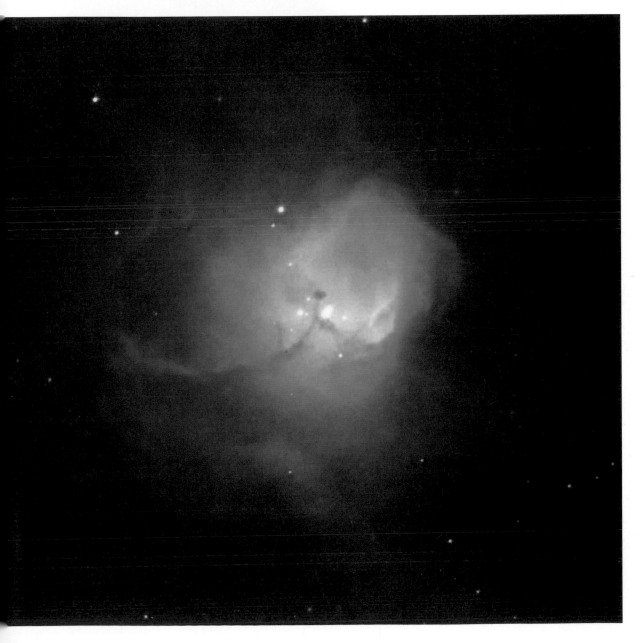

A Universe of galaxies

5

Swirling galaxies are among the most beautiful and evocative objects seen in space, and their study is fundamental to our understanding of how the Universe works. Although galaxies come in a variety of shapes, sizes and masses, they are all essentially huge repositories of millions and billions of stars, plus the clouds of gas and dust from which stars form.

Our own Galaxy, the Milky Way, contains around 100 billion stars, all held together by their mutual pull of gravity. But our Galaxy is not alone. In the observable Universe, there are around 100 billion galaxies. Of these, some are relatively quiescent, but others are tremendously powerful, participating in violent and disruptive galactic collisions.

Galaxies also contain concentrations of "dark matter." A poorly understood component of the Universe, dark matter may constitute as much as 90% of all mass in the cosmos. Another component of galaxies is massive black holes, which have masses of between one million and one billion Suns. Black holes reside in the centers of some (or possibly all) galaxies. They are thought to be the powerhouses for the extremely luminous but distant galaxies called quasars.

Galaxies represent the building blocks of the cosmos. Astronomers are studying galaxies intensely to understand the birth of stars, how galaxies evolve, and the nature of the Universe on the largest scales. Although it has long been known that galaxies are often found in groups called clusters, a stunning revelation of the last two decades has been that these clusters fall into even larger groupings, known as superclusters, which form the largest known structures in the Universe.

▶ *The Andromeda Galaxy, M31, is the largest member of the Local Group. It is the farthest object that can be seen with the naked eye. Andromeda is thought to be a near twin of the Milky Way.*

Galaxies are useful for tracing the history of the Universe because they can be seen to such great distances. Recall that light travels through space at the incredibly fast but finite speed of 300,000 km/s (186,000 miles/s). Galaxies can be seen to distances of billions of light-years away, which means that the light from the most distant observed galaxies was emitted at a time when the Universe was billions of years younger. By observing galaxies at different distances, astronomers peer into a historical record of the development of the Universe, in much the way that archaeologists dig deeper into the ground to find relics from increasingly ancient civilizations. However, before delving into what galaxies reveal about the cosmos, let us first look at the galaxies themselves.

Island universes

Prior to the middle of the 20th century, it was thought that our Galaxy and the Universe were one and the same. Astronomers looking out into space with increasingly powerful ground-based telescopes saw stars, clusters of stars, and a variety of nebulae. Some of these nebulae appeared as fuzzy patches in the sky, whereas others possessed a more regular appearance, being rounded in shape or having a spiral form.

In the early part of the 20th century, there were differences of opinion about the nature of some nebulae, whether they were components of the Milky Way or in fact separate galaxies in their own right. In the spring of 1920, at the National Academy of Sciences in Washington, D.C., there was a debate between two American astronomers, Harlow Shapley and Heber Curtis. Shapley asserted that the nebulae in question belonged to the Milky Way. Curtis, however, claimed that they were distinct from our Galaxy. At the time Shapley was considered to be the "victor" of the debate, but it is now known that he was wrong, for there are literally billions of distinct galaxies. The main problem at that time was that distances for the nebulae and the extent of the Milky Way were not known very accurately. Distances to astronomical objects have been, and continue to be, quite difficult to measure definitively.

The Curtis–Shapley debate serves as a good reminder that scientists draw conclusions from the interpretation of data using a variety of assumptions. Either the data or the assumptions can sometimes be misleading. What led Shapley to the wrong conclusion? It was primarily two things. First, the distance to the nearest major galaxy, called the Andromeda or Messier 31 (M31), was estimated on the basis of the brightness of a nova. (A nova is a very luminous explosive event that occurs in a

▲ *The Coma Cluster of galaxies contains more than 1000 galaxies and lies at a distance of 350 million light-years. The two brightest galaxies seen in this Hubble Space Telescope image are an elliptical (left) and a spiral type (right). The elliptical galaxy is about 15 times more luminous than the spiral. Note that all of the other objects (except for a few foreground stars) are galaxies at much greater distances. One particularly interesting object can be seen at lower middle. It appears to be two galaxies running into each other. The Coma Cluster illustrates the amazing range and form of galaxies in the Universe.*

binary star.) However, the nova under consideration was actually a supernova, which is considerably more powerful, and such objects were not known to exist in Shapley's time. This mistaken identification led astronomers to think that Andromeda was closer than it is now known to be.

Second, Shapley had used a special class of stars called Cepheids to determine the extent of the Milky Way, which he found to be so large that even at a great distance the Andromeda could well be associated with our Galaxy. However, Shapley did not know about the effects of the gas and dust that make up the interstellar medium, which, somewhat like a fog, makes stars appear dimmer than they should. This dimming effect led Shapley to overestimate the distances of the Cepheid stars and thus the size of the Milky Way. So, Shapley actually came to the proper conclusion that the Andromeda was in no way special from other nebulae based on what he knew at the time, but as sometimes happens in science, new observations and advances in technology lead to improvements of the data and the assumptions, which in turn can greatly alter prior conclusions.

In 1925, using a combination of more reliable estimates for the distances to stars plus the discovery of Cepheids actually in Andromeda itself, Edwin Hubble, of Mount Wilson, California, placed the distance to Andromeda at around 1 million light-years (about half of modern estimates), proving it to be external to our Galaxy. Of even greater importance, Hubble measured Cepheids in a number of galaxies and found that galaxies at increasingly greater distances appeared to be moving away, or receding, at increasingly higher speeds. Thus the Universe was found to be expanding!

By 1936 Hubble had determined a value for the so-called Hubble Constant, which relates galaxy distance to speed of recession and has implications for the age of the Universe. Although much has changed since the early 1900s, it is clear that galaxies have had, and continue to have, profound implications for our understanding of the Universe, and are very interesting in their own right in their diversity and evolution.

In a galaxy far, far away...

The Universe is everywhere "sprinkled" with galaxies. To look at galaxies at ever greater distances is tantamount to seeing galaxies as they were earlier and earlier in the history of the Universe. Thus galaxies can be used as markers of how the Universe has changed over time. Since astronomers are keenly interested in how the Universe has evolved, including galaxy formation, finding distances (and thereby ages) of far-flung galaxies is of great importance. Unfortunately, measuring distances to astronomical objects is a major challenge. As has been mentioned before, it is difficult to tell from measurements of only the apparent brightness of a star or galaxy whether the object is truly very luminous and far away or whether it is shining only relatively faintly but located nearby.

So how do astronomers measure distances to galaxies? One way is to find a "standard candle" – a kind of "light bulb" of known wattage, an object of known luminosity that can be used to gauge

▼ *This majestic galaxy, M100, is similar to our own. It is composed of gas clouds and brightly glowing stars. M100 is just one of the approximately 2500 galaxies that can be found in the Virgo Cluster. Mixed among these billions of stars are the pulsating variable stars known as Cepheids.*

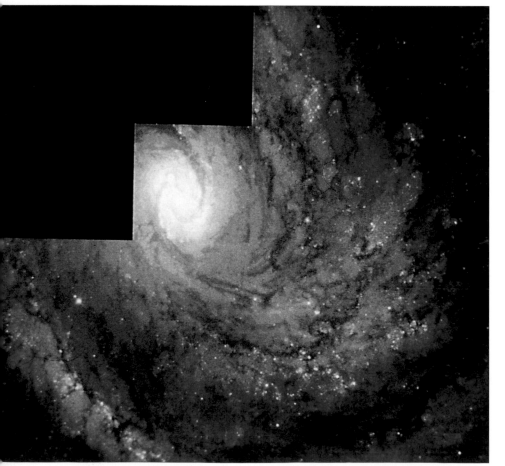

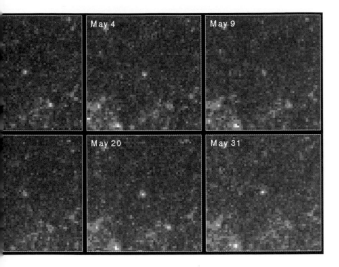

May 4   May 9

May 20   May 31

distances. For example, throughout the 20th century Shapley, Hubble and others used members of a class of variable supergiant stars called Cepheids to measure distances to nearby galaxies. By observing the period of brightness variation, they knew how luminous these Cepheid stars should be, so by measuring the apparent brightness of the Cepheid stars, they could determine how far away the stars were. Cepheids are still used today as a prime means of obtaining astronomical distances.

Although Cepheid stars are relatively luminous objects, they can only be used to measure distances to the nearest galaxies, out to about 125 million light-years, with the aid of the Hubble

Space Telescope. Something else must be used for galaxies that are farther away. One kind of "star" that is used is not actually a star at all but an event – a supernova.

Supernovae represent intense and terminal stellar explosions. A certain class of supernovae, called Type Ia, involves a binary star comprising a normal star and a white dwarf star. The white dwarf is drawing gas from the normal star to itself. There comes a critical moment when enough gas has fallen on to the white dwarf to make it blow up. What makes this event useful for measuring galaxy distances is that the "critical moment" is the same for every binary of this kind, meaning that Type Ia supernovae are good standard candles. Furthermore, supernovae are much brighter than Cepheids, so they can be seen to much greater distances. The Hubble Telescope has identified supernovae out to nearly 10 billion light-years away. At these huge distances, astronomers are viewing the Universe when it was about half its present age.

There are other methods for determining distances to galaxies. Some of them are quite technical and sophisticated, involving, for example, the brightest stars or the brightest galaxies. Currently, however, supernovae are being used to measure distances to galaxies several billions of light-years

◀ *This series of Hubble Telescope images has captured the brightening and dimming of an individual Cepheid star in M100. The regular cycles of Cepheid stars allow astronomers to measure their luminosity and thereby their distance. The Telescope was trained on to a region of M100 for 12 exposures spread over two months. The result was the discovery of 20 Cepheids. Based on these observations, astronomers estimate the distance of the galaxy M100 to be 56 million light-years.*

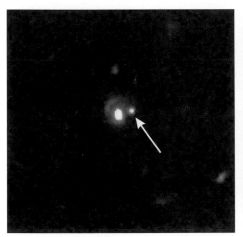

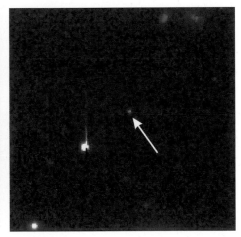

▲ *These distant Type Ia supernovae were captured with the Hubble Space Telescope. The three stellar explosions (indicated by arrows) occurred in 1997 and were initially discovered with ground-based telescopes. The ones at left and middle are at a distance of 5 billion light-years. Both the supernovae (bright points) and the host galaxies (brightish "blobs") are seen. The supernova on right is almost 8 billion light-years away, making it one of the most distant supernovae ever discovered. This supernova is so far away that its host galaxy cannot be discerned in this image. At this distance the Universe is seen at a time when it was only half its present age.*

away, with some rather startling results that have a bearing on the nature of the Universe.

Where do galaxies come from?

Traditional ideas for the formation of galaxies bear strong resemblance to those for the formation of stars and of planets. The picture is not entirely clear, and many astronomers are working hard to understand how galaxies originally formed and evolved over time. The Universe in its infancy was filled mostly with hydrogen and helium gas. In some regions the density of gas managed to be greater than that of its immediate surroundings. The slightly overdense clouds had stronger gravity and therefore attracted more gas from around them. As gas accumulated in these overdense regions, the clouds became increasingly massive, and they started to collapse in size eventually to form galaxies.

At this point the story becomes a little uncertain. Little cloudlets condense out of the huge collapsing cloud to form stars. If these stars form fast enough, then stars are scattered everywhere throughout a roughly spherical or oval-shaped galaxy. These are the elliptical galaxies that will be discussed in more detail later in this chapter. These galaxies have used up all of the original cloud of gas to form stars, so that the ellipticals of today have very little gas in them, which is what astronomers observe. On the other hand, if stars form slowly, then the large gas cloud manages to collapse to a flattened disk-shaped galaxy. These disk galaxies are spiral galaxies with graceful spiral arms. The spirals still have lots of gas and are actively forming new stars today.

This is only a summary sketch of galaxy formation. Astronomers are still sorting out many of the details to understand how, when and why galaxies formed, and, equally of interest, how galaxies have changed over the history of the Universe. Significant advances have been made in recent years, many of them from the Hubble Space Telescope. One of these discoveries is that collisions, mergers and other interactions between galaxies may be important in understanding the different types of galaxies that can be seen. It seems that astronomers have only recently started to unravel the deep secrets of galaxies.

Age of the Universe

▶ *Shown here is a sequence of galaxy pictures with time, with the oldest at left and the youngest at right. At left are examples of the roundish ellipticals and the graceful spirals as they appear today (at a rough age of 14 billion years old, which corresponds to the age of the Universe). The columns to the right display examples taken with the Hubble Space Telescope of galaxies at 9 billion, 5 billion and only 2 billion years old.*

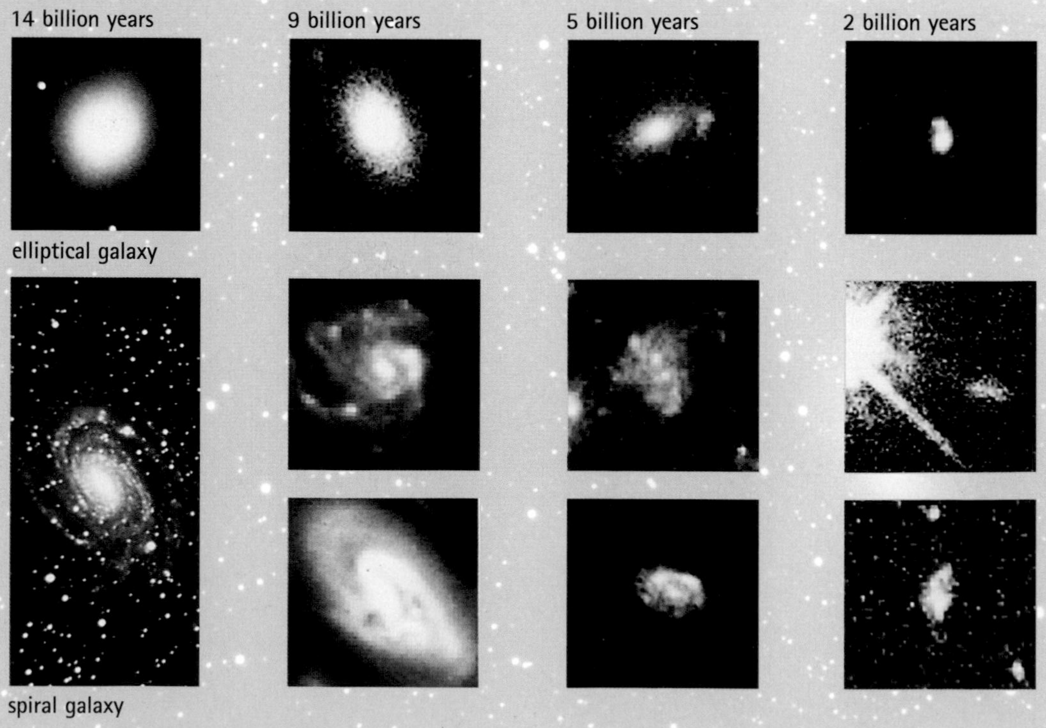

14 billion years

9 billion years

5 billion years

2 billion years

elliptical galaxy

spiral galaxy

The Milky Way

Our Galaxy is called the Milky Way. The name comes from the fact that it appears to trace out a faintly glowing ribbon across the sky. This "milky" band, which can be seen on any clear night, appears so because our Galaxy is shaped rather like a flattened pancake. Since we live inside the Milky Way, our view through the Galaxy is obscured, but looking up or down from within presents a relatively unobstructed view of the Universe. It is similar to being in a shallow ground fog – street lamps that are a block away may be severely dimmed, whereas a full moon overhead can be seen with clarity. So it is when viewing the Milky Way from within, which is where the Earth, Sun and Solar System reside.

Living in the Milky Way makes it both the easiest and the hardest galaxy for astronomers to study. It is easy because everything in it is relatively close by. On the other hand, it is difficult, and sometimes impossible, to view distant parts of the Milky Way through the accumulated clouds of gas and dust that reside in the "pancake" of our Galaxy. Even so, astronomers have made great progress in understanding the different components that make up the Milky Way. We shall see, however, that a few mysteries remain for astronomers to unlock.

Home sweet home

Our own Sun is just one of about 100 billion stars in the Milky Way. The Solar System lies at a distance of some 26,000 light-years from the center

This schematic view shows the Milky Way as seen from above (top) and from the side (below). The spiral arms are confined to a flattened disk, which contains most of the stars and gas. An extended roughly spherical halo holds the globular clusters and most of the dark matter. Our Sun and planets are in the disk, about halfway between the Galactic center and the disk edge.

of our Galaxy. So, just as the Earth is not at the center of the Solar System, the Solar System is also not at the middle of the Galaxy. On the whole, the Milky Way has a luminosity equal to 20 billion Suns and a total mass of about 1000 billion Suns. Being a typical spiral galaxy, the stars and mass are distributed throughout three distinct components: a flattened disk with spiral arms; a central bulge region that is round in appearance; and an extended halo that stretches out to great distances.

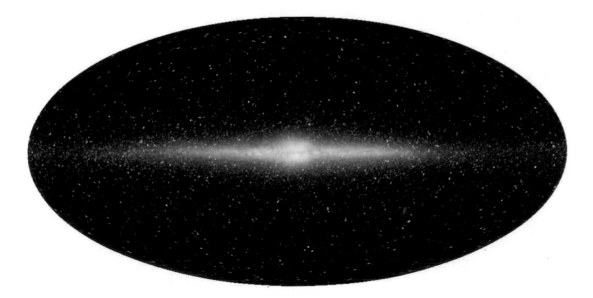

This picture of the Milky Way was taken in infrared light from a space satellite called COBE by looking over the entire sky. It beautifully shows the thin disk of the Milky Way and the central ball-shaped bulge. The disk component is filled with gas and dust. The dust shows up as a reddish color in this image and the stars appear white.

► The image above right
shows the field of stars in
the direction of the center
of the Galaxy, in the
constellation of Sagittarius.
Here, the Hubble Space
Telescope was peering into
a small hole that is
relatively free of obscuring
dust to capture this
impressive display. The field
is extremely dense with
stars because the
perspective is through the
disk of the Milky Way, which
is where most stars reside.
The view shown below right
of the Galaxy center was
obtained using a radio
telescope. The dust and gas
clouds are transparent at
radio wavelengths, thus
allowing for a clear view of
the Milky Way center. The
arrows point out prominent
features. The white arrow
marks Sgr A, the Galactic
center itself, which is about
100 light-years across. The
red arrows indicate Sgr B1
and Sgr B2, which are
regions with actively
forming stars. The yellow
arrows point to supernova
remnants (SNR), indicating
the bubbles of gaseous
debris that have been
ejected by a past supernova
explosion.

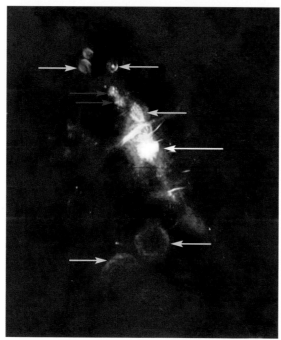

The disk and spiral arms

Most of the stars in our Galaxy lie in the disk component, which is a flattened region roughly 160,000 light-years in diameter but only about 1000 light-years thick. So the disk is more than 100 times longer across than it is in depth. For comparison, coins range from only 10 to perhaps 50 times greater in diameter than in thickness, so the Galactic disk is exceedingly thin indeed. The bulk of gaseous clouds are also in the disk. Much of this gas traces out a spiral arm pattern; the Sun and our Solar System dwell in one such arm.

The disk is rotating, with the Sun orbiting the Galaxy at a speed of 220 km/s (140 miles/s). At a distance of about 26,000 light-years from the center of the Galaxy, the Sun takes some 200 million years to complete one orbit. Since the Sun is around 4.5 billion years of age, it has circumnavigated the Milky Way only about 20 times, such is the vastness of our Galaxy. Another interesting fact is that in spite of there being a few hundred billion stars in the disk, the Galaxy consists mostly of empty space. In fact, if our Sun were to be shot through the disk like a pinball, from one side to the other, the odds of it striking another star before coming out the other end are one in a billion.

A distinctive feature of the spiral arms is that they possess most of the star population in the Milky Way. The arms are rich in gas, which is what forms new stars, and are therefore rich in young stars. They are especially rich in clusters of stars called open or galactic clusters, which typically consist of hundreds to thousands of stars. The brightest stars are the hot, massive, blue ones that eventually explode as supernovae. Although they are far outnumbered by lower-mass yellow and red stars, nuclear fusion occurs at such stupendous rates in these massive stars that they are extremely luminous. Such stars dominate their surroundings and the overall visual appearance of spiral galaxies.

The bulge and central black hole

The Galactic bulge marks the central region of the Milky Way. The bulge is an essentially spherical region about 30,000 light-years in diameter. It is densely packed with primarily lower mass, cool stars, which give the bulge an overall reddish hue. Its center is located in the direction of the constellation Sagittarius. Astronomers are not able to view the centermost portions of the Galactic nucleus directly, at least not in the visible light that we see with our eyes. The space between us and the Galactic center is densely filled with stars, and clouds of gas and dust obscure our view in that direction. Nonetheless, the Galactic center can be and has been studied in other kinds of light, such as X-rays, infrared and especially radio. These observations have revealed several interesting facts about the nucleus of the Milky Way.

An important discovery is that the Galactic center is a source of high energy X-rays, and that this emission originates from a central volume of only 3 light-years in diameter. (Recall that the nearest star to the Sun is over 4 light-years away.) Radio studies reveal that the clouds and stars in this general vicinity are orbiting some central object or objects at large speeds of several hundreds of kilometers per second. Under the reasonable assumption that the motions are governed by gravity, the total mass of this interior region must equal millions of Suns.

Now, for a galaxy with a mass of a thousand billion Suns, a few million does not sound like much, but this few million at the Galactic center is crammed into a very small volume, comparable to our Solar System. It is now widely thought that the X-ray emission and high orbital speeds of stars and gas can only be explained if there exists a black hole with the mass of a few million Suns at the Galactic center. The concept of massive black holes in galactic nuclei is relatively common, and the idea will be taken up again (see page 101) in the context of quasars, for which the black holes are having truly spectacular consequences.

The halo and dark matter

The halo region of our Galaxy is an extended sphere-like volume that stretches out to perhaps 300,000 light-years from the Galactic center. Here dwells a much older population of stars. Although individual halo stars are scattered throughout the halo, there are also nearly 150 known globular clusters, which are dense concen-

trations of this older stellar population. Each cluster can have between 100,000 and 1 million stars, which, owing to their great age, are quite red, in contrast to the much bluer stars found in the young clusters of the spiral arms. Globulars are extremely interesting because they represent a time of star formation when the Milky Way itself was very young and still in the process of forming. Thus, the halo stars are some of the oldest objects in our Galaxy and can be used to estimate the Milky Way's age, in the sense that a galaxy must be older than what it contains. This age works out to about 10–12 billion years.

The halo is also host to a mysterious component of the Milky Way called dark matter. A growing body of evidence has convinced astronomers of the existence of copious amounts of non-luminous mass in our Galaxy and throughout the Universe. Estimates indicate that as much as 90% of all the mass in the Milky Way is dark matter. In other words, there exists a major component of our Galaxy that cannot be directly seen but instead indirectly inferred from its influence on motions of stars and clouds of gas.

The exact nature of this dark matter remains unclear and is a major outstanding mystery of astronomy. Some of this dark matter may come in the form of dead stars, such as cool white dwarfs, neutron stars, or failed stars called brown dwarfs. It seems that some of the dark matter must, however, stem from strange particles smaller than an atom, such as axions or neutrinos. At this time no one is quite sure of the exact nature of dark matter.

◄ *NASA's Chandra satellite captured this X-ray view of the Milky Way center. Thirty separate exposures were combined to make an image that is 400 by 900 light-years along its sides. The colors represent different temperatures of hot gas, ranging from roughly 2 million degrees (red) to 5 million degrees (green) and up to about 8 million degrees (blue). The supermassive black hole at the Galactic center resides in the central whitish patch. The various point sources are exotic stellar remnants (white dwarfs, neutron stars and black holes) and distant galaxies. Nearly four days of satellite observing time were required to produce this X-ray masterpiece.*

► *Some of our Galaxy's oldest stars can be seen in this stunning picture. This cluster is known as M80 (or NGC 6093), and it lies at a distance of 28,000 light-years. M80 is a prime example of the nearly 150 known globular clusters that orbit around the Milky Way in the halo region of the Galaxy. The globulars are symmetric in appearance. They represent sites of the first stars to have formed in the Milky Way. As the oldest denizens of our locale in the Universe, astronomers can use the globular cluster stars to estimate the age of our Galaxy.*

A Galaxy in motion – taken for a spin

Some of the most convincing evidence of dark matter in the Milky Way comes from observing the motions of stars and gas rotating around in the disk. The stars and gas are orbiting under the influence of gravity from the Galaxy as a whole. Now, the gravity that we experience every day is well understood, so that by observing the motion of objects acted on by gravity, astronomers can infer the amount of matter or mass responsible for those motions. The prediction is that the orbital speed should be slower for stars and gaseous clouds located at larger distances from the center of the Milky Way, in much the same way as Pluto's orbital speed is slower than Jupiter's, which is slower than the Earth's. But, incredibly, this does not appear to be the case for the Milky Way. From about where the Sun is and beyond, everything seems to be orbiting around the Galaxy at the same speed. This can only occur if there is more matter in the Milky Way than is accounted for by observed quantities of luminous matter, namely the stars and gas clouds themselves. There is in fact between 10 and 100 times more mass in the Galaxy than is actually seen. This extra material is called dark matter. The Milky Way is not alone as a repository of dark matter, and the motions of stars and gas in other galaxies is one method used in "weighing" it.

A Universe of galaxies

We live in a Galaxy full of stars within a Universe full of galaxies. It is estimated that for every star in the Milky Way, there is likewise a galaxy in the observable Universe, each with billions and billions of stars of its own. We look here at the main kinds of galaxy types and their activity, from the relatively "normal" galaxies, to the spectacular starburst galaxies, and finally the intriguing quasars. Also of great interest are the effects of gravitational lensing in the Universe, where Albert Einstein's ideas about how gravity can actually "bend" light are being observed by astronomers.

From dwarf spheroidals to giant ellipticals

There is a tremendous variety of shapes, sizes and types among galaxies. Edwin Hubble, who established the distances to the nearest galaxies and thereby their distinctiveness from the Milky Way, was the first to attempt a classification of galaxy types. He based his classification on the overall appearance of galaxies, which he neatly grouped in just a few basic categories – irregulars, spirals and ellipticals. Somewhat more recent has been the discovery of several dwarf spheroidal-type galaxies, which tend to be the smallest and least massive of galaxies. Ellipticals as a class include some of the most extensive and massive galaxies observed. The very fact that galaxies come in different shapes raises a number of interesting questions about what makes one galaxy different from another. Before addressing these important questions, the overall properties of the different galaxy classes need to be summarized.

Dwarf spheroidal galaxies

In some ways the dwarf spheroidals are a relatively new class to astronomical study. Although there are several dwarf spheroidal types closely located to the Milky Way, the dwarfs as a class tend to possess relatively few stars and what stars are present tend to be rather loosely spread throughout their volume. This means that these galaxies are faint and are therefore challenging even to find.

Dwarf galaxies are deficient of hydrogen gas and so are not currently forming many new stars. Most of their existing stars were born in the distant past, akin to the ancient stars in globular clusters. Somewhat mysterious has been the discovery that the dark matter content is higher in the dwarf spheroidals than in other galaxies, with as much as 99% of the galactic mass in dark matter form in contrast to 90% for the Milky Way. It is not yet clear why or how dark matter has come to be so concentrated in these otherwise puny galaxies.

Spiral galaxies

Spirals, like the Milky Way, are truly the most beautiful galaxies. Their flattened disk shapes are sprinkled with brightly glowing open clusters of young stars, and their elegant spiral arms form sweeping patterns. It is in the spiral arms of these galaxies that star formation is actively occurring. The arms, therefore, contain young, hot, blue stars, most of which will eventually terminate in supernova explosions. In fact these powerful supernova explosions may be partially responsible for triggering the collapse of giant gaseous clouds, eventually spawning the next generation of stars. So as long as there are substantial quantities of gas, the manufacture of stars seems to be a self-sustaining process. Spiral galaxies also possess halos, which contain globular clusters and the mysterious dark matter.

Note that the spiral arms are not solid features but are actually a kind of wave, shaped like the spiral ripple that results from stirring a can of paint. The spiral arms are seen because the stars and gas orbiting the galaxy become bunched up in this pattern. The stars and clouds do not always stay in spiral arms, but instead pass through them. However, matter congregates in the arms because a star will move slower while in the arm and faster when traveling between the arms. This is not unlike the knots of traffic that develop on the motorway in busy parts of town – cars tend to become backed up where there are many junctions, but zip along on the open motorway where traffic volume is low.

As a class, the spirals represent about 20% of all galaxies. The spiral arms usually extend into the central portions of the galaxies where they can no longer be seen because of the bright bulge region. In some cases (about one in five), the bulges have elongated "bar" shapes, and the spiral arms appear to originate from the ends of the bar. Like the spiral arms themselves, these bars in the galactic nucleus are not solid features, but the appearance results because the stars congregate to make that shape. Our own Milky Way seems to possess just such a bar.

How galaxies compare

(relative to our own spiral galaxy, the Milky Way)

property	elliptical	spiral	irregular
range in mass	0.0001–50	0.005–2	0.0005–0.15
range in diameter	0.01–5	0.2–1.5	0.05–0.25
range in luminosity	0.00005–5	0.005–10	0.00005–0.1

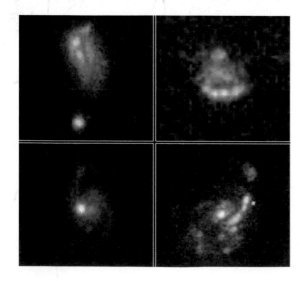

Shown here are different galaxy types. The Antlia Galaxy (top left) is a dwarf spheroidal type. This galaxy is the loose collection of stars that can be seen at center. Although not especially impressive, dwarf galaxies such as this one may be far more numerous than the more prominent large spiral and elliptical galaxies. NGC 1365 (top right) is a barred spiral galaxy. This majestic object is at a distance of 60 million light-years. It is a member of the Fornax Cluster, which is a group of galaxies located in the southern hemisphere. The "bar," which crosses the galaxy center almost horizontally in this image, gives the spiral arms an appearance of originating from "shoulders." Another spiral galaxy is shown below right. It is oriented so as to be seen with a side view. It is known as the Sombrero Galaxy (M104) because of the prominent dust lane that looks like the brim of a large Mexican hat. The giant elliptical galaxy (middle left) known as M87 is found near the center of the Virgo Cluster. This galaxy is four times more luminous and 40 times more massive than the Milky Way. Whereas our Galaxy possesses perhaps 200 globular clusters, M87 has an extensive system of globular clusters numbering maybe as many as 20,000 in total. Bottom left are four distant galaxies imaged by the Hubble Space Telescope's Medium Deep Sky survey. These galaxies are all irregular or peculiar in shape. The bluish hue indicates regions where new stars have just formed.

Elliptical galaxies

The elliptical galaxies range from spherical in shape to something like squashed balls. They have a large spread in luminosity, size and mass. Some ellipticals, called dwarf ellipticals, are only 1 million Suns in luminosity (paltry by galactic standards), but others can range to more than 1 million billion Suns in luminous output. The ellipticals appear to be the most common class of galaxy, representing about 60% of the galactic population. The faint dwarf ellipticals may represent the dominant population of galaxies in the Universe, but astronomers cannot be sure since these galaxies are difficult to detect, just as dwarf spheroidals are hard to find.

In striking contrast to the spirals, elliptical galaxies are severely lacking in gas and dust. They possess few if any young stars and are instead filled with low-mass stars. Elliptical galaxies are dominated in color and brightness by red giants, which have evolved from low-mass stars like the Sun and which eventually end up as white dwarf embers. However, like the spirals, elliptical galaxies do possess globular clusters and also substantial quantities of dark matter.

Irregular galaxies

Galaxies are relegated to this last class when they fail to conform to any of the other simple morphologies. As the name implies, the irregular galaxies do not possess any dominant symmetry in their shape and are in a broad sense lacking in organization of appearance. The Large and Small Magellanic Clouds are two examples of nearby irregular galaxies. Some irregulars, like the Magellanic Clouds, do show evidence of spiral structure but without a clear pattern. The remainder are essentially just a mess of gas and stars, often a consequence of galaxy collisions.

Galaxy collisions are not like those when two cars collide and actually make physical contact. Instead, the stars in galaxies are so widely separated that when two galaxies collide or interact, they essentially pass right through each other. However, if the two galaxies originally had nice regular shapes (like two spheroidals for example), then a collision will lead to an overall irregular appearance for both galaxies. This happens because the gravitational influence of every star acting on every other star as the galaxies pass through one another can substantially distort the original shapes. These giant interacting star systems are sometimes designated as "peculiar," especially if a pair of otherwise regular galaxies are clearly undergoing a collision event. In some cases the collisions are catastrophic such that the two individual galaxies actually merge to become a single new galaxy.

The Local Group – a family portrait

Galaxies are rarely found in isolation, and indeed our own Galaxy is just part of a rather small cluster of galaxies called the Local Group. At present about three dozen members are known in the Local Group, but there are probably more. It is difficult to make a definitive census of the Local Group because some galaxies are exceedingly faint and difficult to find, and some can be hidden behind the disk of the Milky Way, where our view is severely restricted due to obscuring gas and dust. Of the known members, only four are spiral galaxies, including the Milky Way. The rest fall in the elliptical and irregular classes in roughly equal proportions. The most prominent members from our perspective are the Milky Way galaxy, the nearby Magellanic Clouds, and the somewhat farther but significant spiral galaxy known as the Andromeda.

The Large and Small Magellanic Clouds

There are a handful of smaller satellite galaxies that orbit around the Milky Way, very similar to the way that moons orbit the planets. Of these, the two most prominent are the Large and Small Magellanic Clouds (often shortened to LMC and SMC), which can be seen from the southern hemisphere, toward the constellations of Dorado/Mensa and Tucana, respectively. They are named after the Portuguese explorer Ferdinand Magellan, who in 1519 was the first to attempt to sail around the world. The LMC lies at a distance of about 170,000 light-years and the SMC at around 200,000 light-years. They take about a billion years to complete just one circuit around the Milky Way.

◀▼ *At about 200,000 light-years' distance, the Large and Small Magellanic Clouds (below and left, respectively) are relatively nearby traveling companions to our Milky Way. As can be seen, both galaxies are irregular types with a general "unorganized" appearance. In the LMC, the bright purplish-red nebula at center and left is the Tarantula Nebula, also known as 30 Doradus. This region has recently undergone a phase of forming stars at a rapid rate. Some 200 of these stars will explode as bright supernovae over the next few million years.*

Both of these galaxies are quite unlike the Milky Way in several respects. They do not have spiral arms, and they are irregular in appearance. The LMC is only about 23,000 light-years across and the SMC is slightly less than half that, in contrast to 300,000 light-years across for the entire Milky Way. Both galaxies contain less than 1% of the Milky Way's mass. So the neighboring LMC and SMC galaxies are much smaller in size, mass and number of stars, than is ours. But they are also different in another way, providing clues as to their origin and evolution. The LMC has a lower metal content than does the Milky Way, and the SMC's is even lower.

Astronomers oddly refer to "metals" as the relative quantity of elements that are not hydrogen or helium. For example, the Sun is about 70% hydrogen by mass, 28% helium, and the remaining 2% accounts for all other elements, such as carbon, nitrogen, oxygen, and so on. This compositional mix of the Sun is rather typical of other normal stars and gas clouds throughout the Milky Way. For the LMC and SMC, however, the metal content of stars is far less, by factors of between two and ten.

What does the metal content (or "metallicity") tell astronomers about stars? Stars make helium out of hydrogen from nuclear fusion in their cores. In the more massive stars, helium can be fused to make other elements, such as carbon and so on. When a massive star explodes as a supernova, the metal-rich gas of the star is spewed out into its environs. This gas then becomes part of a new star-formation cycle, and the process repeats, with the result that the metal content of the gas gets larger and larger. The present metal content of stars thus relates to the number of cycles and the extent of star formation that have occurred in a galaxy. If the amount of metals builds up gradually over time, then metallicity becomes a kind of archaeological tool in the hands of astronomers, helping them to fathom the historical development of a galaxy.

Another interesting discovery in relation to the LMC and SMC is that the Milky Way is slowly stripping gaseous material away from these two small galaxies in what is known as the "Magellanic stream." The Milky Way's gravity is bleeding star-forming material away from the Magellanic Clouds, thereby suppressing star formation there and probably contributing to their irregular appearance. It is likely that these same gravitational effects will degrade the orbits of the Magellanic Clouds over a long period. In a few billion years, the LMC will be so close to the Milky Way that the strong gravity of our Galaxy will literally rip the LMC apart, causing the two galaxies to merge.

Andromeda – we have a twin

The Andromeda Galaxy was cataloged as a nebula with the designation Messier 31 (M31) by the French astronomer Charles Messier in the late 1700s. At a distance of about 2.4 million light-years, the Andromeda is the nearest galaxy of comparable size and mass to our own Milky Way. In fact, the Andromeda is very like our Galaxy in many respects: it has a central bright nuclear bulge from which spiral arms originate; it has clusters of young hot stars scattered throughout its flattened

▼ *Thought to be a near twin of the Milky Way, the Andromeda Galaxy is the other major member of the Local Group. It is over 2 million light-years away in the constellation of Andromeda. The many white pinpricks of light scattered about the image are foreground stars that are seen when looking out through the Milky Way to Andromeda. The bright elliptical "blobs" at center bottom and lying just above Andromeda are two satellite galaxies that orbit Andromeda, much like the LMC and SMC do for the Milky Way.*

disk; it has a halo region sprinkled with globular clusters and dominated by dark matter; it even has two nearby smaller satellite galaxies called M32 and M110, which are analogous to our LMC and SMC. If we could somehow travel to Andromeda and gaze back toward the Milky Way, our Galaxy would bear a strong resemblance to the view that we have of the Andromeda.

One interesting discovery from studies of the Andromeda is that it and the Milky Way are orbiting each other, much like two stars in a binary. We may even be on a collision course! The Andromeda galaxy is "hurtling" toward our Galaxy at about 470,000 km/h (290,000 mph). That may seem fast, but at a distance of 2.4 million light-years, a collision would not occur for about 6 billion years. However, the Milky Way and Andromeda will not collide if Andromeda has some sideways motion, a motion that is extremely difficult to measure because of its great distance.

Nonetheless, if a collision were to occur at sometime in the future, the two galaxies would likely undergo an epoch of renewed star-forming activity, known as a starburst. Gas and stars might also be channeled into the two respective galactic centers to feed the massive black holes that are thought to be dwelling in quiescence at this time. The collision could even be catastrophic, in the sense that a complete merger of the two galaxies could result. If a merger occurred, the resultant new galaxy would probably lose its spiral identity to become, instead, a more rounded galaxy without a flattened disk, resembling an elliptical.

The spectacular starburst galaxies

The starburst galaxies were originally discovered as extremely luminous galaxies in the infrared region of the electromagnetic spectrum. Typical galaxies such as our own are normally brightest in visible light, which we can see with our eyes. The starburst galaxies are bright in the infrared because they are experiencing a dramatic phase of star formation, and many of the newly born stars are enshrouded in natal clouds of gas and dust. Strong ultraviolet and optical light emanates from these young stars and shines into the obscuring

clouds. The clouds absorb the light and become warm, shining "heat light" in the infrared.

The Large Magellanic Cloud is not a starburst galaxy, but it does possess a microcosm of starburst-like behavior in the region called 30 Doradus (or the Tarantula Nebula). The inner core of this nebula has been intensively observed with the Hubble Space Telescope. Prior to the much better spatial resolving power of modern telescopes, this compact star cluster (known as

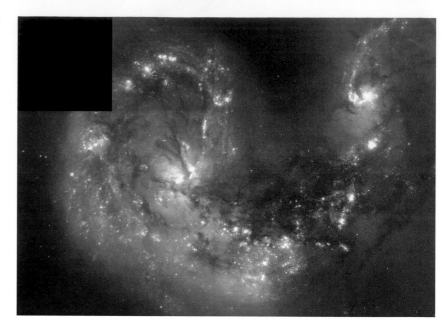

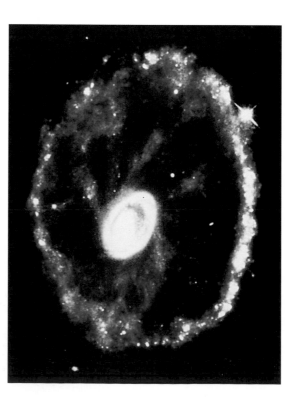

▲ *The Hubble Space Telescope zooms on the central portion of the Antennae, where two galaxies are in the process of merging. The bluish regions are thousands of newly formed clusters of hot massive stars. The bright yellow-orange regions are the centers of the two galaxies. The dark strips are dust clouds lying just in front of our view.*

◄ *The Cartwheel Galaxy appears like a giant blue ring around a yellow core. It was a normal spiral galaxy before it experienced a collision with another fast-moving galaxy (not shown). The ring is huge, with a diameter of 150,000 light-years. It is blue because, like the Antennae galaxies, the collision has initiated a new era of forming young stars.*

R136) was at one point thought to be a single star of around 1000 solar masses. The cluster is now understood to contain some 200 young massive stars, all with masses exceeding 10 Suns. The implication is that every one of these massive stars will eventually explode as a supernova over the period of 3 million years or so (a rate of about one every 10,000 years). It may be that the proximity of the Milky Way to the LMC has, through the effects of gravity, initiated the starburst activity within the R136 star cluster.

It now seems clear that at least some, if not most, starburst behavior is associated with galaxies that have collided in some way. Although the stars in colliding galaxies do not themselves "hit" in the way that two colliding cars do, the huge clouds of gas do make contact. The result is the compression of colliding gas clouds, leading to higher densities and temperatures. It is thought that this compression stimulates cloud collapse, ultimately leading to the starburst activity that is seen today.

What is the sequence of events when a starburst takes place? It seems clear that rapid star formation starts in multiple clusters of stars. One impressive example is the Antennae, where two galaxies are clearly in the process of merging. The Antennae possess as many as 1000 bright clusters of young stars, all or most of them presumably initiated by the merging event. Dominating the light output of these clusters are massive bluish stars. These stellar powerhouses form much more quickly than do low-mass stars like the Sun, so that the clusters quickly "ignite" to shine brightly like hot blue jewels.

Interestingly, massive stars initiate core hydrogen burning so rapidly that the cloudlet in which they form will in some cases not even have finished collapsing. So although just an infant star, these high-mass objects glow with the illumination of 100,000 Suns while still embedded in a dense shroud of gas and dust. Even though these are now stars, they are still in the process of forming since they continue to draw gaseous material from the cloud. The most massive stars live such incredibly short lives, that they can expire in dramatic supernova explosions before stars like the Sun have been born.

Starburst activity as generated from merging galaxies may play a key role in unravelling the mysteries of galaxy evolution. From studies of distant faint galaxies, it now seems that the spiral galaxies were more common in times past than they are today, and that the ellipticals were generally somewhat smaller. It has also been observed that irregular or peculiar galaxies, those that lack organized or uniform shapes, were also more numerous in the past. All of this seems to suggest that galaxy collisions and mergers occurred more frequently in the early Universe. This is reasonable, because with the Universe being in expansion, galaxies would have been much closer together in the past, increasing the likelihood of near encounters.

The prevailing idea now is that many of the early spiral galaxies underwent collisions that resulted in irregular shaped oddities and some

◀ NGC 4650A is one of the few galaxies to have a polar ring, comprising the debris left from an incredible galaxy collision in the past. A remnant of one of the galaxies is the rotating flattened bright galaxy with long side from left to right. The polar ring is the rather ragged, vertical structure that appears to wrap around the more yellow central galaxy. The ring is seen nearly edge-on. The bluish motes in the ring are regions containing luminous young stars. This is yet another example of how galaxy collisions can stimulate star formation.

ellipticals. In some cases, perhaps as many as four to ten spiral galaxies may have merged over a period of time to build up the large elliptical galaxies that are observed in some rich galactic clusters. As spirals merged, the mixing of gas resulted in starburst activity, and the gravitation-al influence of the stars and dark matter destroyed the elegant spiral patterns of the origi-nal galaxies.

These ideas fit with observations by astronomers that today ellipticals are more prevalent than spirals. However, mergers cannot

The Hubble Deep Field

One of the monumental and key achievements of the Hubble Space Telescope has been the acquisition of the Hubble Deep Field (HDF) images, which are rich with information about galaxies in the early Universe. There are actual-ly two sets of HDF images, one from the north-ern hemisphere and one from the southern. The first set came from HST staring at a single small spot (about the size of an average adult seen from a distance of 5 km) in the direction of the Plow (or Big Dipper) constellation. Hubble was trained on this spot for ten full days in December 1995. This "keyhole view" appears relatively vacant in the night sky. It was carefully selected for the very reason that it is devoid of bright stars and nearby galaxies so that the Hubble Telescope could detect the faintest ever seen galaxies.

What did astronomers find? The HDF images contain some 1500 galaxies, most of them about 4 billion times fainter than can be seen with the human eye. Because looking at objects that are far away means seeing them as they were in the distant past, the galaxies in the HDF represent the faintest, farthest and youngest galaxies ever observed. Some of them may be only 1 billion years old, which is less than 10% of the present age of the Universe. The HDF galaxies come in an array of types, from spiral galaxies to ellipticals. But one exciting discovery is that many of the galaxies have peculiar shapes, being neither the majestic spirals nor the rounded ellipticals. Instead, this abundance of irregularly shaped galaxies suggests that mergers and galaxy interactions were relatively common in the early Universe and may therefore have been important in how galaxies have changed since they were formed. The HDF images, from both north and south, will continue to be examined as astronomers strive to piece together the puzzle of galaxy genesis.

▼ *Left and right show the stunning Hubble Deep Fields, North and South. Both images represent a combination of several hundred Hubble Telescope exposures to reveal the faintest and most distant galaxies ever observed. There are several thousand galaxies to be found in these images. At a distance of 12 billion light-years, some of the galaxies are seen at a time when the Universe was only 1 or 2 billion years old.*

Quasars, possibly the most extreme objects in the Universe

In the 1960s, astronomers began to discover a class of radio sources (objects that were bright at radio wavelengths) that when viewed with traditional optical telescopes appeared as star-like points. These "quasi-stellar" objects are today called quasars, and they have presented some of the most challenging problems to modern astronomy.

Most quasars lie at enormous distances of thousands of millions of light-years, and they are among the most distant objects known in the Universe. They are incredibly luminous, being up to 10,000 times brighter than our own Milky Way, and easily 10 times brighter than the brightest giant elliptical galaxies. A typical quasar luminosity equates to a billion billion Suns! (The Milky Way's luminosity is only about 40 million billion Suns.)

▲ This 100,000th exposure of the Hubble Space Telescope shows a quasar at a distance of around 9 billion light-years. The quasar is the bright object at center. To the right is a foreground star in our Galaxy. The star appears as bright as the quasar but is in fact about a million times closer to us. Interestingly, the faint spot just above the quasar in this image is an elliptical galaxy, at a distance of about 7 billion light-years, that just happens to lie in the same direction as the quasar.

explain the existence of all elliptical galaxies, especially the dwarf ellipticals, since even the smallest spiral galaxies are larger than these diminutive objects.

Starburst behavior also occurs where there seems to be no obvious evidence of galaxy interaction and such activity is not as well understood. Our own Milky Way is a spiral galaxy, suggesting that it has remained relatively unscathed from any past interactions with galaxies. Yet the Milky Way may have undergone a phase of starburst activity in its past. Could an epoch of starburst behavior not related to the merging of galaxies be the norm for most galaxies, a kind of galactic puberty? It is thought that supernova events, which are a normal occurrence in galaxies, may stimulate the formation of new stars and might possibly be the source of some starburst behavior even when there has been no collision of galaxies.

▼ There is now excellent evidence that the "engines" of active galaxies are black holes. Part of the evidence comes from powerful jet-like sprays of matter connected to the accretion disk that feeds the massive black hole in the galactic center. For example, this picture shows the active galaxy NGC 4261. At left is a composite image, with the central white fuzzy region being the galaxy as seen in visible light by a ground-based telescope.

Superposed are the jets as seen in radio maps (shown in orange). These radio lobes are of galactic proportions, being about 88,000 light-years in length. At right is a Hubble Space Telescope image of the galaxy center, showing a bright central spot surrounded by a disk. A massive black hole with a mass exceeding 1 million Suns is thought to reside at the disk center, which is also the source of the huge radio lobes.

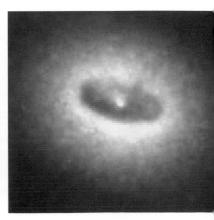

Even more startling is the fact that this extreme luminosity emerges from a relatively compact region. Quasars are observed to vary in light output over a time period ranging from hours, days or weeks to months. Because light travels at a finite speed, the time variation can be related to the size of the emitting object, which falls in the range of 10 to 100,000 astronomical units. Recalling that the astronomical unit (AU) is the distance from the Earth to the Sun, 10 AU is about the size of Saturn's orbit, and 100,000 AU is only a third of the distance to the nearest star. Thus, astronomers have been driven to the conclusion that quasars are objects in which the energy output of many galaxies is crammed into a volume of only roughly Solar System proportions.

It is widely thought that the only explanation for these remarkable properties lies with supermassive black holes. If there is material in the vicinity of these black holes, and some mechanism for getting this material to fall into the black hole, then a tremendous amount of potential energy is available for conversion into emitted light. But what is potential energy? Imagine ascending a tower and on reaching the top, releasing a ball. The ball falls to the ground because of gravity. When the ball impacts the ground, it has a speed and therefore a kinetic (or "traveling") energy. But of course when the ball was dropped, it had no speed at all, so where has this traveling energy come from? It came from having been carried up the stairs. The ball acquired potential energy when it was lifted the height of the tower. Dropping the ball released the potential energy inherent in the ball's elevated position as kinetic energy.

For quasars, instead of the ground there is the supermassive black hole. Instead of a ball, there are clouds of gas or even whole stars orbiting around this black hole. The clouds and stars are at some distance from the black hole, and therefore have a potential energy, much like the ball being some height off the ground. These clouds and stars can be made to fall toward the black hole, releasing their potential energy as kinetic energy.

The infall of gas probably occurs through a structure called an accretion disk, which surrounds the black hole. As the name implies, an accretion disk is like a flattened pancake of gas in orbit around the central black hole. What is being gathered inward is the gas, which moves through the disk eventually to be consumed by the black hole. As the gas accretes, it heats up. The incredible quasar luminosities come not from the black hole (because it is "black") but from the surrounding accretion disk. The part of the disk nearest the black hole is both hot and energetic, radiating powerfully in many regions of the electromagnetic spectrum, including X-rays and ultraviolet.

The amount of light generated in such a scenario depends primarily on the mass of the black hole and the amount of material falling toward it. Black hole masses of between 1 million and 1 billion Suns are required to explain the observed quasar luminosities. This is equivalent to the entire mass of a star cluster or a small galaxy! The gas infall rates can be up to one solar mass every year. This is not to suggest that a single star like the Sun falls into the black hole once every year, instead gas continuously dribbles in, with an average rate of one Sun per year.

Several scientists have attempted alternative explanations for the quasar luminosities, invoking for example dense clusters of bright stars and supernovae or other compact objects that are not black holes. None of these explanations appears to satisfy all the constraints of the energy budget and compact size as naturally as do black holes.

There are other important pieces of evidence that support the black hole and accretion disk model. One is that orbital speeds in the disk should be increasingly large for gaseous material closer to the black hole. This prediction has been tested with observations by the Hubble Space Telescope for the nearest active galactic nuclei (AGN).

The AGN are the broader classification for galaxy nuclei that display high luminosities from compact volumes. Quasars represent the more powerful members of this class. In the nearby AGN, the Hubble Space Telescope has shown that accretion disks do exist and that they do have high rotation speeds. In fact, it is these rotation speeds combined with the theory of gravity that allow one to derive the high masses of the central black holes. Another important observation in support of the accretion disk and black hole model is the occurrence of amazing bipolar jets that originate at the AGN and extend for hundreds of thousands of light-years in length, easily rivaling the sizes of most galaxies. It seems that these jets can only be explained within the black hole paradigm, combined with strong magnetic fields in the vicinity of the black hole.

Active galaxies – just a matter of perspective

There seems to be a wide variety of energetic active galaxies, such as the extremely luminous quasars, or the Seyfert galaxies with bright nuclei, or the stunning double-lobed radio galaxies such as Centaurus A. Some emit strongly in X-rays and some do not. Some vary extremely rapidly with time and some do not. Most have radio emission, but some are brighter at the radio wavelengths than others. Can black holes with accretion disks explain all of these observations? Currently, the answer seems to be "perhaps." Notice that if something is spherical, like a ball, it looks the same no matter how you hold it – it is round. But a disk is different. Take for example a table plate. Setting the plate on the table and looking down on it, the plate appears round. But if you stoop to look at the plate from the side, then you see its edge. Just as with the plate, so it is with an accretion disk. The appearance of the disk depends on the viewing perspective. This single fact alone (with additional technical sophistications) seems adequate for understanding the variety of active galaxies seen by astronomers. As a specific example, the active galaxies that are X-ray bright are the ones with disks viewed more face-on, like a table plate when viewed from above. In this way the middle portion of the disk where the X-rays come from is exposed. The active galaxies that are X-ray faint are the ones seen more side-on or edge-on. In this case the middle parts cannot be seen so easily and so neither can the X-rays be detected. It is the constant challenge of astronomers to understand the many different and sometimes bewildering observations of various objects, and active galaxies certainly represent a major triumph in this endeavor.

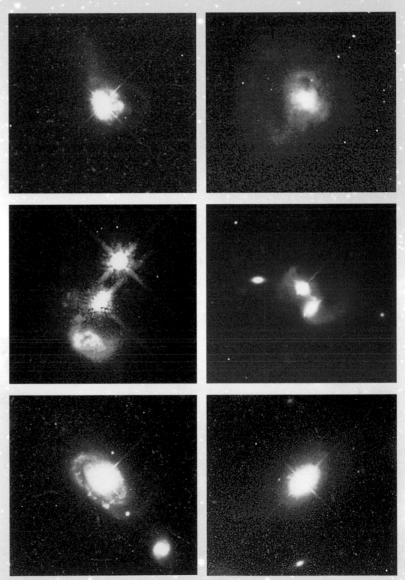

▲ Quasars appear to be "housed" in a wide variety of galaxies. At bottom are quasars in what appear to be typical spiral and elliptical galaxies. At middle are quasars associated with colliding galaxies. The uppermost bright object in the left-center picture is a foreground star, whereas the other two objects are galaxies that have collided at about 1.5 million km/h (about 1 million mph). At top are quasars in peculiar or irregular galaxies. The left one shows a wispy tail, possible evidence of a recent galaxy interaction; the right image appears to show a merger in progress. Note that all of these quasars are between 1 and 3 billion light-years away.

So, after three decades of intensive theoretical and observational study, the gross properties of AGN seem to be understood at a basic level. By definition, a black hole can never be directly observed, yet the presence of a black hole has consequences, through its strong gravitational field, on its environment. Telltale signs of a black hole's presence include high orbital speeds of stars and gas, hot X-ray emitting disks, and narrow energetic jets. Yet, there remain several outstanding problems. In particular, quasars appear to have been more common in the past than they are today, so how do quasars turn on? And did they turn off? What becomes of quasars after they have turned off? Are all galaxies seen today a remnant quasar of yesterday? Astronomers are seeking to address these very interesting questions.

Gravitational lensing – a trick of light

The theory of gravity first advanced by Sir Isaac Newton in the mid-17th century has been incredibly successful in describing a whole host of physical observations, ranging from falling apples to colliding galaxies. Newton's theory implied the existence of black holes, for which there now seems ample evidence in the form of massive companions in some binary stars and as the central engines of the incredible quasars. However, Albert Einstein's theory of gravitation, commonly known as "The Theory of General Relativity," developed in the early 1900s, was

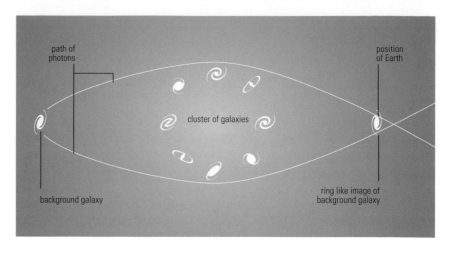

needed to describe accurately the many complicated effects associated with the intense gravity near white dwarfs, neutron stars and black holes. Einstein's theory holds one more trick up its sleeve, and it happens that astronomy, and particularly galaxies, provides one of the best backdrops for it to show up. This effect is the bending of light in the presence of mass, better known as gravitational lensing.

General relativity makes some truly astounding, and sometimes very non-intuitive, predictions. Among its claims are that mass and energy are equivalent, and that mass curves space. It is this latter property that is of concern for gravitational lensing, because light travels through space. Recall the analogy of a taut rubber sheet distorted by a heavy rock that was used in Chapter 4 to describe the curving of space by a massive body. Rolling a ball on this distorted surface deflected the ball's trajectory from a simple straight line. This is how the path of light is modified in the presence of mass. In fact if light passes especially near a black hole, not only will its path be severely skewed, but also, having entered the black hole, the light can never escape. It is lost forever.

Why do we not observe this bending of light at the Earth on a daily basis? It turns out that the deflection of light from a straight trajectory is extremely small except when in the presence of a large concentration of mass. In fact, the first verification of this rather stunning idea (considered by many to be preposterous) came in 1919 when Sir Arthur Eddington promoted an expedition into the southern hemisphere to detect the bending of

▼ Well over 50 years ago, the famous physicist Albert Einstein proposed a theory of general relativity, with one of its predictions being that gravitational lensing would be possible. Shown

here is a rare quadruple lens in the shape of a cross. The four blue spots are the multiple images of a single distant galaxy. The lens is the central red object, an elliptical galaxy that lies between us and the distant blue galaxy. Because lensing is an effect of gravity, observations like this can be used to probe the dark matter content of the lens, in this case the halo of the elliptical galaxy.

▲ This cartoon schematic illustrates how gravity "bends" or focuses light like a lens. A distant galaxy (shown at far left) shines light in all directions. If only the Earth and that galaxy existed in the Universe, then some of the light would come directly to us and the galaxy would be seen where it truly is. But if there are massive bodies lying directly between us and the galaxy (in this example, a cluster of galaxies), some of the light that would normally miss the Earth and not be seen is now "bent around" the massive bodies and can be observed. In the schematic, two rays of light (shown in yellow) come from around either side of the cluster of galaxies, so that at Earth, the galaxy that was the distant light source now appears to be in two different places. Amazing as this sounds, examples of these powerful gravitational lenses in nature are being found in space.

starlight as it passed near the Sun during a solar eclipse. The Sun is the most massive object in the Solar System, so the bending of starlight, although small, was just measurable at that time. The total solar eclipse was vital to the discovery because the lensing effect could only be seen for stars very nearly lined up with the limb or apparent edge of the Sun. Except during a solar eclipse, the Sun is far too bright to observe these stars with the eye. The quest was a total success: the amount of lensing was exactly that predicted by Einstein's theory.

The subject of lensing was then quietly tucked away as an observational study until late in the 1970s when double images of quasars were first observed. The lensing of light from quasars is similar to that of distant starlight passing near the solar limb, only in the case of quasars the light was emitted billions of years ago and it has been lensed not by a star, but by a galaxy or even a whole cluster of galaxies.

How is it known that the light has been lensed, in contrast to there just being a double or binary quasar? There are two telltale signs. First, gravitational lensing can produce multiple distorted images of a distant background source. The images may be distorted and no two look the same, yet the light stems from the same object for all images. So if the quasar varies in some way, perhaps dimming or brightening slightly, that variation will be seen in each of the multiple images (although perhaps not all at the same time). Second, if the alignment of the background

◄ This is a rather complicated example of the lensing of a distant galaxy by an intervening cluster of galaxies. The cluster galaxies are yellow in this image and consist of ellipticals and spirals at a distance of around 5 billion light-years. The object being lensed is nearer to 10 billion light-years away, and it is seen here as bluish arcs. These arcs are the multiple and distorted images produced as a result of the lensing. The number of images, their shape and placement all depend on how the mass of the lens is spread out in space. Each of the cluster galaxies contributes to the lensing, so the situation here is quite complicated. Imagine trying to read using a magnifying glass that is distorted, perhaps with some holes or air pockets in the glass, and with the lens thick in some places but thin in others. The effect produced by such an unusual magnifying glass is akin to the effect that this cluster has on the images of the far away blue galaxy.

quasar and foreground galaxy or cluster is favorable, the multiple images can be smeared out into an arc or even a ring. Such shapes are obviously distinct from galaxies and clusters and serve as clear indications of lensing.

Lensing is an important tool for astronomers in several ways, but possibly one of the most important is for "weighing" dark matter. The number of lensed images and their apparent position depends on the mass of the lens and how that matter is distributed within the lensing galaxy or cluster of galaxies.

Recall that evidence for dark matter comes from the rotation curves of galaxies. There are other evidences of dark matter, but the point here is that this component, although "dark," does reveal itself through its gravitational influence on its surroundings. So dark matter too can lens light. Studies of quasar lensing systems have been used to deduce the amount of dark matter in galaxies and clusters. The results are consistent with other methods in showing that between 10 and 100 times more matter comes in this mysterious form than in luminous stars and gas. The lensing also permits the distribution of dark matter to be mapped. In cases where this can be done in clusters and galaxies, the dark matter is seen to be coincident with the galaxies and more concentrated toward the cluster centers.

A Universe of voids and walls

Our journey through the cosmos has gradually been progressing from the small to the increasingly large, here to culminate in the largest structures observed. We have seen that there are moons around planets, planets orbiting stars, concentrations of stars in galaxies. But what are the largest structures in the Universe? Galaxies also form clusters or groupings. In fact it is rare to find a single galaxy in isolation. But beyond even the clusters of galaxies, the clusters themselves appear to collect in filamentary like structures in the Universe called superclusters, which can span tens and hundreds of millions of light-years.

Clusters of galaxies

Although galaxy types have been discussed on an individual basis, most galaxies dwell in clusters. Our Milky Way is part of the Local Group, which is a relatively small cluster of just a few dozen galaxies. However, the nearby Virgo Cluster, at a distance of 55 million light-years, is a rich cluster of around 2500 galaxies.

The richest clusters typically contain thousands of galaxies. Although the galaxies of such a cluster may be spread over a volume of perhaps 10 million light-years, many of the galaxies tend to be concentrated toward the center of the cluster. The majority of members in a rich cluster are ellipticals, and there is always at least one giant elliptical (sometimes referred to as a "cD" galaxy) near the cluster center. It is clear that with such a dense collection of galaxies in the cluster, collisions and mergers will be frequent, thus accounting for the many ellipticals and especially for the extremely large cD galaxies, which are thought to arise from multiple merger events.

Superclusters – tracing out the grand design

As will be seen in Chapter 6 on cosmology, the Universe is thought to be isotropic and homogeneous when considered on very large scales. Isotropy refers to how the Universe appears in dif-

▼ This Hubble Space Telescope image shows a distant cluster of galaxies, 8 billion light-years away. About 80 galaxies can be identified within the cluster. Of these, around a dozen show evidence of recent galaxy interactions, the most ever found in a single cluster. It seems clear that dynamic and sometimes very violent activity was the norm for galaxies in the early Universe.

ferent directions, whereas homogeneity refers to how the Universe appears from different places. Imagine standing in a dense forest of pine trees, so that the forest looks the same in every direction. From where you stood, the forest would be isotropic in appearance. Now if the Earth were everywhere covered in pine trees, then the forest would also be homogenous, with every patch of forest essentially the same as every other patch. If, however, the forest did not cover the Earth, then it would not be homogeneous, since some bits of the Earth had forest and some bits did not. Nevertheless, it might still appear isotropic if you stood in the midst of the dense pines.

The Universe is thought to be characterized as isotropic and homogeneous, yet in mapping out the positions of galaxies to as far as several hundred million light-years, there appear to be large regions bereft of galaxies ("voids") and other regions where the galaxies fall along elongated structures ("filaments"). Some of these voids can be 50 to 150 thousand light-years across. (Note that the "voids" are not true vacuums; they do possess some gas and galaxies, but relatively little when compared to the filaments.) In contrast to the immense voids, there exists a huge collection of galaxies in the form of a long narrow sheet known as the Great Wall. This enormous "wall" stretches to about 250 by 750 million light-years in size. The Milky Way resides in a somewhat disk-shaped supercluster structure with a diameter of 100 million light-years.

Not only are there walls and voids, but there are also motions. That does not appear too surprising in itself: the Moon moves around the Earth, the Earth around the Sun, the Sun around the Galaxy, and so on. But whereas there are regular motions inside galaxies, there should not be any preferred motions in the Universe as a whole. The galaxies are moving, but they should not move in any special direction. Interestingly, there does appear to be a preferred direction in our little "corner" of the Universe.

Astronomers have discovered that the Local Group and other nearby galaxy clusters are all moving in a particular direction, toward some massive but unseen object named the "Great Attractor." This "attractor" lies at a distance of

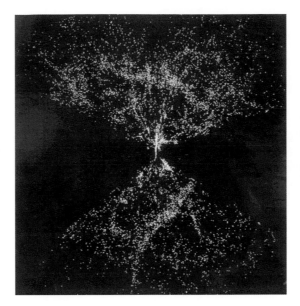

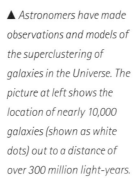

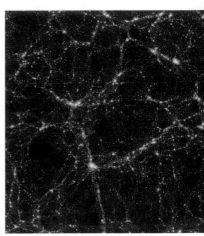

▲ Astronomers have made observations and models of the superclustering of galaxies in the Universe. The picture at left shows the location of nearly 10,000 galaxies (shown as white dots) out to a distance of over 300 million light-years.

The Earth is at center. The picture represents a cut through a large survey of the whole sky. It is rather like cutting out a slice of watermelon and noting the distance and direction of every seed from the center. The Great Wall is seen as a

collection of about 1000 galaxies strung out across the middle of the top half. The collection of galaxies falling along a diagonal in the lower half of the image is called the Southern Wall. "Voids" are the darker regions lacking galaxies (that is, lacking the white dots). The picture at right shows one of the many simulations of galaxy formation created with supercomputers. This "slice," like the slice showing the observed galaxies, is a model for how mass in the Universe clumps over time to produce the voids, walls and filaments traced out by galaxies. The orange filaments represent dense regions of dark matter where galaxies are predicted to form. The darker portions are of low density and represent the voids. For scale, each side of this picture is about 600 million light-years in length.

some 160 million light-years and must have a mass of about 10 million billion Suns (or 10^{16} solar masses). Such a high mass is comparable to the largest known superclusters. There appear to be some galaxies in the region of the Great Attractor, but it is not clear if there are enough to account for the huge mass. Perhaps the Great Attractor is a dense concentration of dark matter. At this time it remains another outstanding mystery in our understanding of how the Universe has evolved.

If the Universe is supposed to be regular throughout, where has all this structure come from? Why are galaxies not uniformly spread throughout all space? How were the voids made? And for that matter, how were the filaments consisting of galaxies made? Although astronomers continue to puzzle over these basic questions, they think that the answers are intimately linked with the fundamental processes operating in the early Universe, a subject that falls firmly within the dominion of our next topic, cosmology.

The life and times of our Universe

6 At this stage of our journey through the Universe, we find ourselves peering at the cosmos from the broadest of perspectives, and in some ways the grandest of all. Throughout history, people have longed to understand the origin and workings of space. Cosmology is a subject that brings many of these questions to bear on the Universe as a whole.

The term cosmology originates from the Greek to mean literally "the study of order." Today the word is defined as the study of the origin and nature of the Universe. Since the Universe refers to all existing matter, energy and space, there is a real sense in which all subjects of scientific study fall into the domain of cosmology. However, in practical

usage, cosmology refers to how the Universe has come to be as it is today. Thus, it is not so concerned with stars or even galaxies specifically, but rather with the study of these objects in terms of addressing questions such as: How did the Universe begin? What is the age of the Universe? What will the Universe be like in the future?

This is an exciting time in cosmology. The Hubble Space Telescope and giant telescopes on the ground are capturing views of some of the farthest galaxies, when the Universe was just a fraction of its present age. The Universe is awash in a cool bath of light left over from the infancy

of the Universe, and this radiation is supplying important clues about the conditions at that time and the way in which galaxies would subsequently form. It is also clear today that the mysterious dark matter plays an important role in the formation of galaxies.

One of the most important astronomical discoveries in the last hundred years has been that the Universe is expanding. This discovery was heralded by Edwin Hubble himself, after whom the Space Telescope is named, and it marked the beginning of rapid developments in our physical understanding of the cosmos. Before delving into the most recent results, however, a short historical prelude will help to put the modern view and its achievements in better perspective.

A brief history of cosmological thought

Astronomy seems to have been considered of practical importance in one respect or another by every known ancient culture. Most of these diverse cultures, including the Babylonians, the Chinese, the Egyptians and the Mayans to name a few, concentrated on making careful observations of the motions of heavenly objects. For example,

they would diligently note the rising and setting times and positions of the Sun, planets and prominent stars. From such records, stretching back over decades and even centuries, these different civilizations devised calendars to mark seasons and special events. They were also concerned with predicting eclipses of the Sun and the Moon.

It would appear, however, that these cultures were primarily concerned with the predictive power of the sky and had little interest in understanding why planets and other celestial bodies move in the way they do. Although the Babylonians and others kept meticulous observational records of the sky – still an important aspect of astronomy today – it is generally the geometry-minded Greeks who are credited with having first developed physical models to explain celestial motions. (It may be that much Greek thought originated with the Egyptians, but this is unclear because of the scarcity of ancient Egyptian records.)

The culmination of Greek cosmological thought was essentially geocentrically oriented, meaning that they took the Earth to be at the center of the Universe. They thought that the Sun, Moon, planets and stars all revolved about the Earth. The Greek astronomer Aristarchus, however, was an outstanding exception. Around 270 BC, he advocated a cosmology that was heliocentric, one in which the Earth rotated about its axis and orbited around the Sun, which is exactly the modern understanding of our Solar System. Aristarchus even estimated the size of the Earth, Moon and Sun and also the distances to the Moon and Sun. For various reasons, his ideas were rejected by his peers, and it was not until the early 16th century that the Polish astronomer Copernicus developed, or reinvented, the heliocentric description of the Solar System.

Copernicus' heliocentric view was adopted by the 17th-century German astronomer Johannes Kepler, who was trying to understand the motion of Mars in the sky. Against the backdrop of the constellations, Mars appeared to execute a loop on the sky over the course of a couple of years. Other planets, such as Jupiter, also do this,

◀ *This false-color image of the entire sky was made by the COBE satellite using infrared light. The plane of our Galaxy can be seen clearly across the middle of the image. The infrared light emitted by the cool stars defines the Galaxy's central bulge.*

but the loops are not so pronounced. Kepler, who possessed an especially good set of observations of Mars made by the Danish astronomer Tycho Brahe, found that he could only understand the loop pattern of motion if the orbits of the Earth and other planets around the Sun were not exactly circular but instead slightly oval (described as elliptical). Abandoning circular motion was a huge conceptual leap: since the time of the Greeks, circles had been considered "perfect" geometrical shapes that must certainly be capable of explaining the heavenly motions.

Although Kepler could now explain planetary motions, he had not really proved the heliocen-

Olbers' Paradox

Cosmology is viewed as concerning the grand scheme of things. Attacking cosmological problems can sometimes require substantial mathematical skills and the immense computational power of large supercomputers. However, an inquiry into the nature of the Universe can begin at the door step.

In the early 19th century, an astronomer and physician from Vienna by the name of Heinrich Wilhelm Olbers posed a very canny question: "Why is the sky dark at night?" Such a question appears naive at first. It is dark at night because, as everyone knows, the Sun has set. But Olbers' question is actually more interesting. The word "Universe" normally brings a number of images to mind: perhaps it is an expanse filled with galaxies, or perhaps it is a sense of vastness. One way or another, ideas about infinitude come to mind – space stretching out forever in every direction, galaxies of stars and planets that will last for a very long time. Olbers' simple little question actually tells us whether these ideas can be right.

For example, suppose the Universe were infinite in extent and infinite in age. Then if stars were spread everywhere and equally throughout the Universe, the entire sky should be ablaze, being as bright as the surface of the Sun wherever and whenever one looks. If the Universe was filled with an infinite number of stars, there would be no point in the sky where one could look that would not somewhere intercept a star; since a star would therefore be seen at every point in the sky, the sky should be as bright as the stars themselves. Clearly this is not the case, and an explanation is demanded.

Although others had pondered the question of night's darkness, it was Olbers who, during the years 1823–26, brought the matter to attention. The issue has since become known as Olbers' Paradox. It finds a resolution in a few different and profound possibilities. One solution is simply that the Universe does not possess an infinite number of stars. But if this is the case, and the Universe goes on forever, then whatever stars there are must somehow be "huddled" together in some recess of the Universe, with everywhere else being devoid of stars. This idea seems rather odd and unlikely, although technically not impossible. Sometimes in science, and especially in a subject such as cosmology, scientists will allow themselves to be guided by a principle known as Ockham's Razor, which is the premise that the most plausible explanation is the one that is simplest and that contains the fewest assumptions. The idea that in an infinite Universe stars would only be found in one part of it seems rather contrived and, therefore, according to Ockham's Razor, unlikely. Beware, however, that Ockham's Razor is never definitive, but where there is ignorance, it can at least suggest a good place to start.

Another solution is that the Universe could be limited rather than infinite in extent, meaning that there could perhaps be a finite number of stars. This explanation seems equally unsatisfying, if not worse, since the implication is that the Universe has some kind of boundary "edge" in space, which is a somewhat bizarre concept (what would be on the other side?).

Supposing again that the Universe is infinite in both time and space, then perhaps dust (in the interstellar medium) could obscure the distant stars, explaining why the sky is not illuminated everywhere. This is rather a good idea because galaxies are known to possess dust. However, obscuring dust turns out not to resolve Olbers' Paradox because the dust would block the light from distant stars by absorbing it. Having done so, the dust would become warmed and reradiate light. So it remains that the sky will appear to glow in every direction, which of course it does not.

The last possibility for resolving the paradox is that the Universe has changed with time, and the simplest expression of this is that the Universe has a finite age. Since light travels at a finite speed, then even if the Universe is infinite in size, extending forever in every direction, stars can only be seen to a certain distance. So it may be that in any direction one looks, there is a star at the end of that sightline, but the Universe is

tric construct to be reality. It was his contemporary Galileo Galilei who managed to do this. Galileo began looking at the sky with a telescope. In 1610 he observed the phases of Venus, a planet that, like the Moon, displayed full and gibbous phases. These phases should never be seen in a geocentric model, but they were to be expected from the Copernican hypothesis, and thus the central position of the Sun in our Solar System was demonstrated.

The notion that the position of the Sun represented not only the center of the Solar System but also the center of the Universe remained for several hundred years. In fact, as measurements of

simply not old enough for its light to have reached the Earth yet, hence the night sky is dark.

A major achievement of cosmology in the 20th century was the discovery that the Universe does appear to be finite in age, that it had a beginning. This then becomes a natural solution to Olbers' simple yet profound question, one that satisfies Ockham's Razor, and one that you can appreciate whenever you glance at the night sky.

▲ *Center top in the painting* Starry Night Over the Rhone *by Vincent Van Gogh is the familiar constellation of the Plow (also known as the Big Dipper). The stars appear to be suspended against the backdrop of a dark night sky. The fact that night is dark has profound implications for the nature of the Universe, for if the Universe were infinitely old, of infinite extent, and everywhere filled with stars, then the night should shine like the day Sun. Why the night is in fact dark and not bright is known as Olbers' Paradox.*

the distances to nearby stars began to accumulate, it did indeed appear that the Sun was the center of the observed Universe.

In the late 1700s, the English astronomer William Herschel sought to deduce the shape and extent of our Milky Way galaxy by counting up the stars in different parts of the sky. He assumed that stars existed only in the Galaxy and nowhere outside it. Within the Galaxy, he further assumed that stars were uniformly spread out with distance from Earth. Herschel's efforts correctly revealed that the Milky Way was shaped like a flattened disk, but his results incorrectly set the size of the Milky Way at about 6000 light-years in diameter and 1000 light-years thick, with the Sun placed quite near its middle. More than a hundred years later, the Dutch astronomer Jacobus Kapteyn used the same basic method and determined that the Galaxy is larger than was thought, yet with the Sun still somewhat near the center.

That the Sun was at the center of the Milky Way, which at that time constituted the entire Universe, was an erroneous interpretation. It came about because the obscuring dust present between the stars, which greatly diminishes the brightness of distant stars, was not understood at the time. As discussed in Chapter 5, it is this same obscuring dust that led astronomers in the early 1900s to overestimate the size of the Milky Way and to think that the nebulae (actually sep-

Doppler shift

The Doppler shift causes the whistle of an approaching train to be of high pitch, and then to be of low pitch after the train has passed. If the train were motionless, its whistle would be in-between in pitch and unchanging. So for sound, the Doppler shift changes the pitch in a manner that depends on whether the train is coming toward or going away from us. This same principle works for light, but with the Doppler shift changing the color of the light instead of the pitch of the sound. Imagine someone carrying a yellow torch and running toward you. The yellow light would take on a bluish tint. But if that person were running away, the yellow light would then take on a reddish tint. By the way, this person would have to run very fast indeed because a change in color from the Doppler shift effect is quite small unless the light source (like the torch) is traveling very rapidly. The torch would have to be approaching at about 15% of the speed of light, or 48,000 km/s (30,000 miles/s), to change yellow to red. Under the right conditions, however, modern instruments can measure Doppler shifts of a few meters per second (10 km/h or 6 mph) in astronomical objects. Such measurements allow velocities to be calculated.

arate galaxies) were just part of our own Galaxy. There is actually so much dust throughout our Galaxy that this early work at surveying stars in visible light could only pick out those within a distance of 10,000 light-years or so from Earth. It is easy to understand why Herschel and Kapteyn found the Sun to be near the center of the Milky Way – they could only see stars in the nearest 10,000 light-years in every direction.

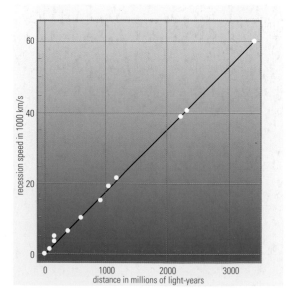

◀ In 1929 Edwin Hubble made measurements of distance and speed of recession for several galaxies. His results led him to the profound discovery that the Universe is expanding. This plot represents a modern example of what is now known as the Hubble Law. Distance is along the horizontal axis in millions of light-years. Along the vertical axis is speed in thousands of kilometers per second. Each dot represents the speed and distance of a single galaxy. For example, the point at upper right is for galaxies in the Hydra Cluster, which lies over 3000 million light-years away and is observed to be traveling away from us at a speed of 61,000 km/s. The points at lower left are all relatively nearby bright galaxies (about 20 in number). The straight line guides the eye in showing that the more distant galaxies are moving at higher speeds. Although few of the points fall exactly on the line, it is evident that all the points follow along it quite closely. The slope of this line (similar to the gradient of a hill) yields the so-called Hubble Constant, which can be used to estimate the present age of the Universe at about 14 billion years old.

The development of cosmology during the past 100 years

The realization that the Sun lies nowhere near the center of our Galaxy came in 1915. The American astronomer Harlow Shapley was using Cepheid variable stars to derive distances to globular clusters. (This is the same Shapley who debated with Heber Curtis on the nature of other galaxies, as discussed in Chapter 5.) Shapley showed that these clusters of stars were peppered about in space in a roughly spherical volume. However, the spread of these clusters in space was not centered on the Sun, but instead on a point in the direction of the constellation of Sagittarius, about 25,000 light-years away.

An appreciation for where the Sun and Earth lie in relation to one another and in relation to our Galaxy advanced rather slowly over the course of human history. By contrast, our understanding of the properties and contents of the Universe has developed spectacularly in only the past 100 years. One of the two major discoveries of immense cosmological significance occurred in 1929 when Edwin Hubble convincingly showed that galaxies were receding from us, and that the more distant a galaxy is, the faster it moves. This has come to be known as Hubble's Law, and it signaled the realization that the Universe is expanding. This discovery was a significant turning point in the study of the Universe, and it marked the beginning of modern cosmology.

Only a decade prior, Albert Einstein had developed his theory of general relativity, which describes how gravity works. He and others, such as Alexander Friedmann (in Russia), Willem de Sitter (in the Netherlands) and Georges Lemaître (in Belgium) to name a few, were now beginning to apply this theory to describe the nature of the Universe at the largest scales. The discovery of the expansion of the Universe came at just the right time to complement the work of these theorists. Hubble's findings set the stage for the development of the Big Bang theory and the subsequent discovery and interpretation of the faint cosmic glow that pervades the Universe.

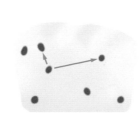
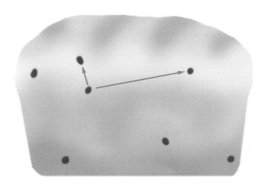

In with a bang!
Hubble's expanding Universe

Edwin Hubble was not in fact the first to show that galaxies were receding from us. Neither was the discovery of an expanding Universe entirely a surprise. In the years 1913 to 1917 the American astronomer Vesto M. Slipher had found that some galaxies were moving away from the Earth at high speeds, but he had not drawn the conclusion that the Universe is expanding. Albert Einstein and others were trying to predict the state of the Universe by using general relativity to describe the effects of gravitation. These scientists found that an expanding Universe was a natural consequence of the theory of general relativity. But it was Hubble's work at the end of the 1920s that put these ideas on a solid observational footing.

What did Hubble actually do? He measured the distances of nearby galaxies, using Cepheid variable stars, and he measured the speed of motion for these galaxies, either toward or away from us, using their Doppler shift. Having accumulated distance and Doppler shift measurements, he found that the two were related. The more distant galaxies were found to be traveling away from the Earth at greater speeds than the closer ones.

Broadly speaking, this observation suggests either that the Milky Way is at the center of the Universe and everything is moving away from it, or that the Milky Way is not in a special location, but the entire Universe is expanding so that all galaxies appear to be receding no matter the location of the observer. Historically, the concept of where the center of the Universe resides has gradually changed, with it moving farther and farther from the Earth. The idea that the Milky Way is at the center of the Universe and that everything is

▲ What does it mean to say that the Universe is expanding? Here an analogy is drawn from the rising of raisin bread when it is baked. Raisins are spread throughout the dough, and as the bread rises it carries the raisins along so that they separate. The key observation is that the separation between any and every pair of raisins is lengthened. For each raisin, therefore, it appears that all the others are drifting away from it, with the farther raisins moving faster. Similarly, it is space itself that (like the dough) is expanding, and galaxies are merely along for the ride. The result is that it seems that galaxies are moving away from us in every direction.

expanding away from our Galaxy seems rather incredible, especially since the Milky Way seems to be a rather typical galaxy. So, the notion that the entire Universe is in expansion is the accepted working hypothesis of modern cosmology. Besides it is also known that the Milky Way is moving toward the Virgo Cluster, which would seem odd if our Galaxy were the true center of the Universe.

But why is the Universe expanding? How did the expansion start? Will it ever stop? Where is it

The distance ladder

The Hubble Law for the expanding Universe is important for several reasons. One of those reasons is that the so-called Hubble Constant, which is the slope of the line that relates galaxy distance to its speed of recession, can be used to estimate the age of the Universe. If astronomers can accurately measure recession speeds and distances to lots of objects, both near and far, then the Hubble Constant can be determined and so can the amount of time that has passed since the Big Bang. It turns out that recession speeds are relatively easy to measure, but gauging distances to the farthest galaxies is much harder. Inferring galaxy distances is a difficult process based on a variety of methods in a hierarchical scheme known as the "distance ladder." Here, just a few of the more important "rungs" will be touched on.

One of the most important techniques used to measure distances to nearby galaxies involves a class of pulsating stars known as Cepheid variables, as discussed in Chapter 5. Because the period of pulsation and consequent light variation of these stars is related to their luminosity or wattage, Hubble was able to use these variable stars to measure the distance to the Andromeda Galaxy as well as to many other nearby galaxies. With the Hubble Space Telescope, Cepheids have been used to work out the distances to galaxies in the nearest big cluster toward Virgo (about 60 million light-years) and beyond.

An important method used to estimate the distances to the farthest galaxies involves the particular class of supernovae known as Type Ia. These supernovae occur in binary star systems. Mass from a normal star is transferred to a white dwarf companion until some critical amount has accumulated that causes it to blow up. There are

▶ *This galaxy, NGC 4603, is one of the most distant galaxies in which Cepheid variable stars have been identified and studied. The Hubble Space Telescope was used to spot over thirty Cepheids, the study of which yielded a distance to the galaxy of 108 million light-years. (Note that the brightest stars – those with "spikes" – are actually foreground stars in our Galaxy and are not related to NGC 4603.) Edwin Hubble's use of Cepheid stars to measure distances to relatively nearby galaxies allowed him to discover that the Universe is expanding. After more than 75 years, astronomers continue to use this "tried and true" method, pushing it to ever greater distances, in order to better understand the nature of the expansion of the Universe.*

two key features of the Type Ia supernovae that make them so useful to astronomers. First, the explosion is extremely bright and can be seen to large distances in the Universe. Second, the critical threshold at which the star blows up is essentially the same in every instance.

Type Ia supernovae, therefore, make excellent "standard candles" because they all output nearly the same wattage. In contrast, some supernovae are the result of a single massive star exploding; such events, although extremely luminous, vary in wattage output for various reasons. The Type Ia supernovae, however, are much more like "light bulbs" all of the same make. These supernovae have been used to measure some truly distant galaxies, perhaps as much as 8 billion light-years away. As we will see later in this chapter, the results are prompting cosmologists to reassess the nature of the expanding Universe.

expanding from and what is it expanding into? These are all important questions, and none is especially easy to answer. But one of the amazing things about cosmology is that steps are being made toward addressing these fundamental issues.

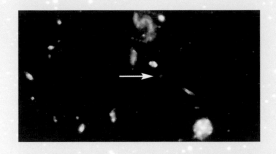

▲▼ *Shown here are images of two distant galaxies. In the image above, the arrow points to a faint galaxy that appears as just a pinprick in this picture. This galaxy is part of the Hubble Deep Field image. It is thought to be one of the most distant known galaxies, even farther away than many of the most distant quasars. If the distance estimate is correct, the light from this galaxy was emitted when the Universe was less than 1 billion years old. The picture below shows a cluster of galaxies that* *has distorted the image of a much more distant galaxy through gravitational lensing. The lensed galaxy is marked by an arrow; it appears as a faint reddish arc. The cluster is estimated to be at a distance of 5 billion light-years, but the lensed galaxy may be up to 13 billion light-years away (assuming the Universe is 14 billion years old). Distant galaxies like these, which existed when the Universe was only a billion years old or less, are challenging astronomers to understand how galaxies formed.*

The cosmological principle

There are several blatant questions about the concept of an expanding Universe that need to be examined. Many of them are addressed with the so-called cosmological principle. This principle states that the Universe is homogeneous and isotropic, which means that, when considered over great distances, every place in the Universe is basically the same as every other place, and that the Universe looks the same in every direction.

As an example of homogeneity, think of a group of 500 children passing through a cafeteria line, all getting a helping of beans and sausages. Sitting down at tables, all of the children manage to swap trays, only to discover that what they had after trading was pretty much what they had before – essentially the same plate, same fork, same serving. Such is homogeneity.

Isotropy is a little different. It means that things appear the same no matter which direction one looks. In this case a counterexample is useful. You are reading these words in a place that is almost certainly not isotropic. If in a room, there is a ceiling, a floor and four walls. Perhaps one wall holds a door, and one wall has a window. Some walls may have pictures whereas others may not. When you look at different walls, you see different things. The room is not isotropic from your vantage point.

How is any of this relevant? Accepting that everything appears to be receding from us, then a natural conclusion to make would be that the Milky Way must surely be at the center of the Universe. But this has often been a mistaken assumption of the past. The cosmological principle would state that there is no center and instead the Universe is everywhere expanding. How does the idea of expansion occurring at all places help to explain why all galaxies seem to be moving away from us?

A common example to illustrate this point is to imagine baking a raisin dough (as in the illustration on page 113). As it bakes, the bread begins to expand. The raisins are not expanding, only the bread. But as the bread expands, the raisins are carried along, so that for any one raisin, every other raisin seems to be moving away. In the case of the Universe, the thing which

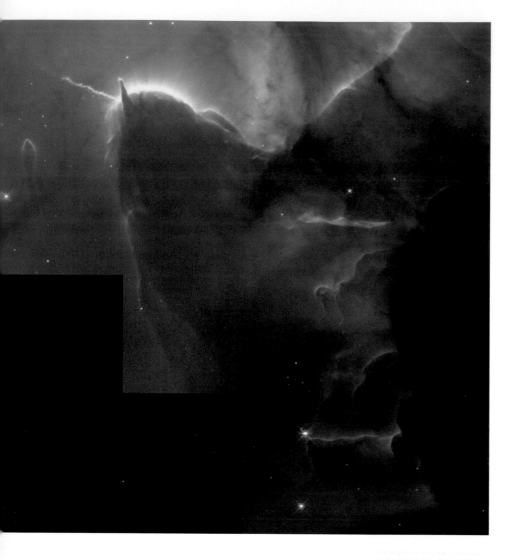

is expanding is space itself (the bread), and the galaxies (the raisins) are just along for the ride. There is no center anywhere, instead the expansion is everywhere.

It is worth pointing out that the cosmological principle is something that is assumed. Is there any evidence to verify it? In terms of isotropy, it does appear that the Universe looks nearly the same in every direction. First, as discussed in the next section, there is a glowing of the Universe that was produced at a very early time, and this glow is not only seen in every direction, but is also almost indistinguishable in different parts of the sky. Second, there are galaxies everywhere astronomers look. One way to test isotropy is to count up the galaxies in a given direction and to compare the result to counts of galaxies in some other direction. If the Universe is isotropic, these counts should have approximately the same number. If one looks far enough, such as in the Hubble Deep Field provided by the Space Telescope, there do appear to be similar numbers of galaxies in every direction. The question of homogeneity is more difficult. It is not possible for us to view the Universe from another vantage point. An astronaut cannot simply jump into a rocket and blast off for a distant galaxy to have a look around!

▲▶ *The stunning image above displays just a small portion of the Trifid Nebula, an immense cloud of gas that nurtures newly forming stars and is located 9000 light-years away. There are many stars in and around this nebula, and there are a great many such nebulae spread throughout the Milky Way. With such a variety of stars and nebulae, it hardly seems likely that the Universe is isotropic, for in our Galaxy every view seems to hold a different treasure. However, drawing an analogy from people,* *although everyone recognizes that no two individuals are alike, when set in a large crowd individuality becomes rather lost against the backdrop of humanity. In the same way, when gazing across vast expanses of space to see many thousands of galaxies, the Universe looks pretty much the same all over. This is demonstrated in the image at right, where over 2 million galaxies, covering a mere one-tenth of the entire sky, are shown. This map is for the part of the sky toward the south*

galactic pole, looking out of the plane of our Galaxy. The blue, green and red colors represent bright, medium and faint galaxies. Black areas represent regions that *have been excluded from the survey. It does appear that galaxies are fairly uniformly spread throughout space, yielding basically identical views in* *whichever direction we may choose to look. This suggests that the Universe is isotropic, which is fundamental to the study of cosmology.*

A faint glow from the distant past

The conception and verification of the expanding Universe naturally implies that at some time long past, everything in the Universe was much more compact and closer together. Actually, everything that is now in the Universe was not in the Universe at its very earliest stages, in the sense that there were no planets, stars or galaxies. Instead, the Universe was incredibly dense and extremely hot, being filled with many different kinds of particles, such as electrons, together with intense radiation.

In the 1940s a Russian-American astrophysicist by the name of George Gamow advocated such ideas for the conditions in the Universe at its earliest stages. That the Universe is seen to expand proportional to distance is reminiscent of explosive events observed in nature. For example, the gaseous layers of a massive star are expelled during a supernova in just this way, at least initially. Consequently, Gamow envisioned the Universe as originating in an explosive "fireball" event; it is this idea that forms the basis of what is now known as the Big Bang theory.

In Gamow's model the early Universe was very hot, thereby resulting in intense radiation, in the same way that a star is brilliant. The Universe, however, was much hotter than any star and therefore much brighter. As the Universe expanded, it also cooled, just as water steam shooting out from a kettle will expand into a room and cool to room temperature. If the present age of the Universe is between 10 and 20 billion years old, Gamow predicted that this ancient remnant glow (known as the cosmic background radiation or CBR) from the initial fireball event should now have a temperature of between 5 and 50 degrees above absolute zero, or around $-270°C$ to $-220°C$ ($-454°F$ to $-364°F$).

In the example of the stars, a hot star has a surface temperature of around 20,000°C (36,000°F) and is predominantly an emitter of ultraviolet light. A somewhat cooler star, like our Sun, has a surface temperature of around 6000°C (11,000°F), and it emits mostly visible light, the band of the electromagnetic spectrum that we can see. Even cooler stars, at around 2500°C (4500°F), are strong infrared sources. The CBR, which turns out to be just three degrees above absolute zero ($-270°C/-454°F$), is bright at millimeter wavelengths, which are like radio waves. This kind of light is not seen by our eyes, but the Earth's atmosphere is somewhat transparent to light in the millimeter band, and so it can be detected, although with some difficulty, using ground-based radio dish telescopes.

A view from afar

What would it be like to live somewhere else – not merely to live on a different planet around some other star in the Milky Way, but actually to live in a different galaxy altogether? To live in an elliptical galaxy, for example, would present a quite different perspective of the night sky. First, there would no longer be any "Milky Way" since ellipticals are not flattened galaxies; instead we would see stars in nearly equal numbers all over the sky. Second, the stars would have a mostly yellow or reddish hue, since there is very little gas to form new stars. For example, there would be no counterpart to the nearby stellar nursery the Orion Nebula; however, there would still be planetary nebulae, because Sun-like stars evolve to become giant stars and "puff" away their outer atmospheres.

Alternatively, instead of a different galaxy, imagine we inhabited a planet in a different cluster of galaxies. The Milky Way is one of the dominant members of the Local Group, but in the Virgo Cluster, which has over 2000 members, ours would be a rather ordinary galaxy. As a rough estimate, the Virgo Cluster has about 2500 galaxies and a diameter of 13 million light-years. The typical separation of galaxies is about 0.5 million light-years, which is around four or five times closer than the Andromeda Galaxy is to the Milky Way. If the Andromeda were this close, it would no longer appear as a faintish blob about 3° across in the sky; instead it would rival the Large Magellanic Cloud in brightness, and it would appear much larger, at 15° across (the angular size of your fist at arm's length). However, the spectacular views afforded by such a location would come with risk: there would be a much greater likelihood of our Galaxy having collided with another some time in the past.

But no matter where we might live, it would only be the details that are different. Cosmologists maintain that whether we live in Andromeda or in Virgo or anywhere else, the Universe as a whole would still have the same overall appearance that we see from our Galaxy. We would still see an expanding Universe, distant quasars, and so on. And using these observations from a different location, we would arrive at the same conclusions as we have at Earth. This is the essence of the cosmological principle.

► Arno Penzias (left) and Robert Wilson (right) are standing in front of the Bell Telephone Laboratories' horn antenna in New Jersey, United States. In 1965, while looking for sources of interference with terrestrial radio communications, these two men discovered the cosmic background radiation emanating through the Universe. Not only did they identify this relic bath of radiation, but they also discovered that it had about the right temperature, being 3 degrees above absolute zero as others had predicted. Their find is historic for confirming the overall picture of the Big Bang theory, and they subsequently received the Nobel Prize for physics in 1978.

Gamow had only predicted this faint glow in the radio band – it was not actually observed until 1965. Astronomers at Princeton University in the United States began constructing a radio telescope to look for this glow, but, at the same time, two sci- entists working for Bell Telephone Laboratories in the nearby state of New Jersey stumbled across this cosmic emission while trying to identify sources of interference with satellite communications. For this discovery, the two men, Arno Penzias and Robert

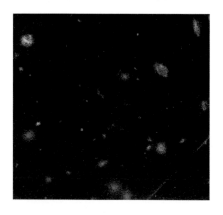
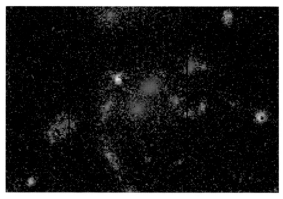
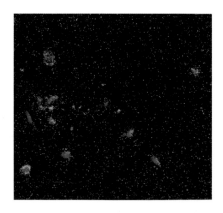

▲ Above are three distant and large clusters of galaxies. The galaxies shown as red are those belonging to a cluster. At left, the blue galaxy is likely a foreground galaxy and unrelated to the more distant cluster. The other galaxies form a group of ellipticals and spirals at about 5 billion light-years away. At center, the galaxy cluster lies at a distance of some 4 billion light-years in the direction of the Andromeda constellation. At right, this cluster lies in the direction of Taurus. It is about twice as distant as the previous two examples and thus very young. Galaxy clusters like these must have developed from regions of the early Universe where matter (in particular dark matter) was slightly more concentrated, as revealed in the cosmic background radiation maps obtained with the COBE satellite.

Wilson, later received the 1978 Nobel Prize for physics. The expanding Universe and the CBR arguably rank as perhaps the two most important discoveries in cosmology to date. They provided convincing support for the Big Bang model.

As predicted, the CBR is observed to have a low temperature, and to come from everywhere in the Universe with equal strength. Since its discovery, the CBR has been studied a fair amount with terrestrial telescopes. However, it became clear that observations of high quality across several bands in the millimeter region would require a space-borne mission. In 1989 a satellite called the Cosmic Background Explorer (COBE) was launched into orbit to observe the CBR over the whole sky, thereby producing a comprehensive map of the radiation. It took several years to survey the sky at different radio bands, but the result has been a stunning confirmation of a theory that was advanced 50 years earlier.

Supernovae at the brink

When Albert Einstein applied his theory of general relativity to the Universe, the fact that it was expanding had not yet been discovered. The thinking of that time was that the Universe should be static, meaning unchanging with time or position in space. To make the theory match his expectation, Einstein had to introduce a new mathematical term into his equations, something he called the cosmological constant. The effect of gravity is to attract matter, so the Universe could not be unchanging since the mass of galaxies would be pulling on each other and inducing motions. The cosmological constant was introduced to oppose that tendency in order that the Universe might remain static. When Hubble discovered that the Universe was expanding, Einstein referred to the cosmological constant as his "biggest blunder."

Recent findings from research teams who hunt for distant supernovae, however, suggest that there is something valid about the idea of the cosmological constant. In the standard Big Bang model, it is thought that the Universe is expanding slowly, with the matter in the Universe causing a deceleration of the expansion because of gravity. However, the most distant supernovae appear to be farther away

COBE

The Cosmic Background Explorer (COBE) satellite map of the cosmic background radiation (CBR) is a true marvel of modern astronomy. The two major results of COBE's survey are that the nature of the CBR is exactly as expected and that it is very nearly the same in every direction. In the first instance, the spectrum of the CBR, which is how the amount of light changes from one wavelength band to the next, is matched by the predicted theoretical curve to a very high degree of accuracy. This is strong evidence that Gamow's fireball model is a good description of the early Universe. In the second instance, the amount of light from the CBR was found to be the same to about one part in 100,000 from different points on the sky (that is, the same amount from the direction of the North Pole as from the South Pole, for example). This means that the CBR is incredibly uniform, and in some ways almost too uniform.

The Universe that we see now is filled with structure, such as galaxies and stars. It is thought that the structure was seeded by tiny fluctuations of density in the early Universe, when matter was much more uniformly spread throughout space. The places of slightly higher density had somewhat stronger gravity. Those places, therefore, attracted more matter, which made the gravity that much stronger, thereby attracting more matter, and so on, until agglomerations of gas formed, which would eventually make galaxies and clusters of galaxies. But such density fluctuations should leave an imprint in the CBR, with some places being brighter and others being fainter. With variations of just one part in 100,000 of the CBR, that imprint is small indeed, but it would appear that the density fluctuations are just large enough to account for the clusters of galaxies that now exist. The resolution to the problem may lie with the enigmatic dark matter, which could greatly increase the density fluctuations without leaving any hint of its presence in terms of the CBR's brightness.

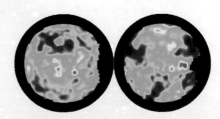

▶ These sky maps of the cosmic background radiation (CBR) were made with the Cosmic Background Explorer (COBE) satellite based on four years of observations. After much processing, the image shows variations of the CBR brightness in false color. At left is the view when looking up from the plane of our Galaxy. At right is the view when looking down in the opposite direction. The red areas are slightly hotter and, therefore, slightly brighter than the blue areas. The brightest and faintest parts vary by only a few parts in 100,000. So, the CBR is extremely regular all across the sky, looking very nearly the same in every direction. The small variations in brightness indicate that matter was slightly clumped in the early Universe, a necessary condition to seed the growth of the galaxies that are seen today.

than they should be, which implies that the expansion of the Universe has actually been accelerating. So it would seem that gravity is insufficient to halt or retard the expansion of the Universe and that there is some other force that acts as a mechanism of repulsion, pushing thing apart and causing space to stretch out faster now than it did in the past. As a generalization of Einstein's cosmological constant, this repulsion effect is referred to as "quintessence," a Greek term referring to a mysterious "fifth element."

But how does quintessence actually work so as to accelerate the expansion of space? No one is entirely sure, and an explanation may involve complex ideas from quantum physics, but it may be that the vacuum of space possesses just a little bit of energy. If so, this energy would translate into a pressure that would seek to expand space as time marches on. Such an effect would be quite small at any one locale but important over large distances in the Universe.

An immediate consequence is that the Universe may be older by a few billion years than was thought. It is clear that more research is needed to determine if there are other explanations for these supernova observations. For example, perhaps the space between galaxies (the intergalactic medium) is filled with some unknown absorbent material. This new source of extinction would make distant supernovae appear faint, and so astronomers would overestimate their distance from us. In the same way, Shapley overestimated distances to Cepheid variable stars because he was not aware of the intervening absorbent dust of the interstellar medium in our Galaxy. Or perhaps these distant supernovae, which occurred when the Universe was much younger, are in some way different from what we know as Type Ia supernovae today. However, if the expansion of the Universe is truly accelerating, and the evidence is proving to be increasingly convincing, then the interpretation of the data may yet invoke new physics. Whatever the eventual result, this is truly an exciting time for cosmologists.

A cosmic calendar in review

Much of this book has described the Universe as it is seen today, with its planets, stars and galaxies, and has given some explanation for the way in which many of these objects have come to be as they are. Now it is time to consider the same approach for the Universe itself. Imagine launching back into time to view the Universe as it was at different moments in its history. The Universe contains countless galaxies and their constituents: How have they come to be that way? What were they like in their embryonic state? What was the Universe like before any galaxies had yet developed? Here a descriptive list of highlights in the cosmic calendar is presented as they are currently understood.

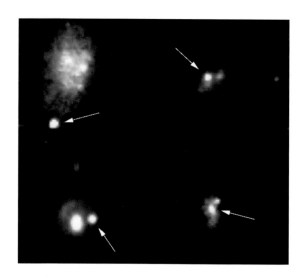

▲ *These distant supernovae have been "caught in the act" by the Hubble Space Telescope. Arrows indicate the supernova explosions. The associated "fuzzy" objects are the host galaxies for these luminous explosions. The four examples shown here range in distance from about 3.5 to 5 billion light-years, meaning that the light from these explosions originated several billion years ago and has only recently arrived at Earth. Distant supernovae like these are proving to be of immense importance for measuring the expansion of the Universe long ago.*

The early Universe

The early Universe was hot and dense. Cosmologists do not really know what was happening during the first moments of the Big Bang, that is, the first 10^{-43} seconds. In this extremely small sliver of time, the conditions of the Universe were

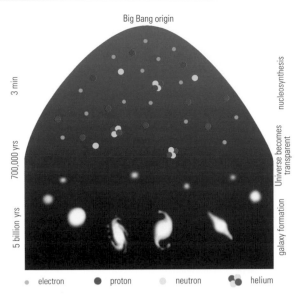

Big Bang origin

nucleosynthesis

Universe becomes transparent

galaxy formation

3 min

700,000 yrs

5 billion yrs

● electron ● proton ○ neutron ●● helium

▶ Major events in the development of the Universe are represented in this schematic. At around 14 billion years ago, the Universe had a beginning in a phenomenally powerful explosion called the Big Bang. As a result, space itself began to expand. The Universe may have undergone a brief period of inflation, when the expansion was incredibly rapid, followed by the relatively passive expansion that has continued until today. The principal constituent of the newborn Universe was light. Matter consisted of protons, neutrons and electrons. After just two minutes, the Universe had expanded and cooled down to a "mere" 100 billion degrees so that bare hydrogen protons and neutrons could combine to make helium. However, it took around 700,000 years before the electrons and protons could combine to make hydrogen atoms. This is when the light now known as the cosmic background radiation originated. Within a billion years, galaxies and clusters of galaxies had formed. Since then, galaxies have become the hosts of the stars and planets seen today. Galaxy formation has largely ceased, whereas stars and planets can still be made. Meanwhile, the Universe continues to expand, perhaps without end.

so extreme (for example, temperatures in excess of 10^{27} degrees) that the theories physicists use to describe things like gravity and electromagnetism do not apply. It is only after this time that "normal" physics is thought to begin operating. Although many different things happened in the first few seconds and minutes, the major events include the creation of matter, the formation of hydrogen and helium atoms, and, somewhat later, the emergence of the cosmic background radiation (CBR).

An immediate result of the initial explosion is that the Universe began to expand, and it has continued to do so since. As space becomes more and more stretched out, the temperature of the Universe (with reference to the waveband of the glow) has decreased. Today the CBR has a temperature of just three degrees above absolute zero (–270°C/–454°F), but it was much hotter in the early Universe, and this had great importance for determining the events that occurred then.

The Universe at its infancy was full of light, or photons, and that is all there was. During the first second, these photons were very energetic, and two photons of light could interact in a process called pair production to make two matter particles. It is at this point that the protons, electrons and neutrons, which are important as the building blocks of atoms, were created. After this period, the temperature was so great that most of these particles were probably zipping around at nearly the speed of light. Perhaps most of the dark matter was also produced in this period. The photons

must have been extremely energetic, and the process of pair production only operated when the Universe was hotter than about six billion degrees, corresponding to the first minute or so.

About 100 seconds after the Big Bang, the temperature in the Universe had dropped to around one billion degrees, much cooler than before but still extremely hot, with a temperature 65 times greater than that at the center of the Sun. One clue about the early Universe comes from the fact that the most abundant element after hydrogen is helium, which accounts for about 25% of all the gaseous mass in the Universe. A temperature of one billion degrees is sufficiently cool to make helium from protons and neutrons via nuclear fusion. At higher temperatures, collisions between atomic particles are so energetic that atoms like helium, with two protons and two neutrons, cannot stay together. Of course, the centers of stars can be hot enough to make helium, but there is far more helium in the Universe today than can have been produced in stars. So, it would seem that most of the helium that now exists must be primordial, which is to say that it must have existed before any stars had been born and was therefore produced in the early Universe.

The gas in the early Universe was entirely ionized, which means that electrons were free to move on their own without being associated with protons or helium atoms. As a result the photons of light would have bounced around between these free electrons, like high-speed rubber balls ricocheting off billiard balls on a snooker table. In this way light and matter interacted with each other. As the Universe cooled to about 3000 degrees, however, the electrons and protons could come together to make normal hydrogen atoms. At this time the Universe was some 700,000 years old. Once the electrons became part of atoms, the light no longer had anything to bounce off, so the Universe simply "shined," the photons traveling unimpeded through space.

It is this ancient light that is now seen as the cosmic background radiation (CBR), and it is telling us about the Universe when it was less than a million years old. For example, the fluctuations in the COBE maps reveal the early "seeds" of what would eventually grow to become galaxies. Importantly, it also means that light cannot be used to "see" the Universe as it was before this period. It would be like driving several miles through a dense fog. After emerging from the fog, one can turn to see where the fog had ended but nothing beyond it can be discerned.

The aid of inflation

The Big Bang model successfully explains a number of observed phenomena: the Universe is expanding; the gas in the Universe comes primarily in the form of hydrogen and helium in the predicted quantities; and the cosmic background radiation (CBR) is seen in every direction and in nearly equal amounts. The power of a model or a theory always rests on its ability to explain what is seen and to make predictions that can be tested.

So the Big Bang theory passes in several areas, but does it fail in any areas?

The answer is yes, in the sense that the Universe has features that the Big Bang model either would not predict or does not explain. Einstein's theory of general relativity indicates that space can be curved, much like the way in which the surface of a ball is curved as compared to the flat surface of a table. The severity or mildness of the curvature of space can be anything, and yet the space in the Universe seems to be extremely flat. Why should that be so? Also, the CBR is extremely uniform, meaning that it is the same everywhere. But how could distant parts of the Universe, parts that are too widely separated for light to have traveled between them, know that they should have the same temperature? Furthermore, the Big Bang model does not answer the question "Why the Big Bang?"

Dates to remember

Much has happened in the roughly 14 billion years that the Universe has existed and only a few major events have been touched on. One way to organize events in history is to place them on a time-line. The ages involved for the Universe are so great, however, that, like many other dimensions in astronomy, they can be difficult to grasp. To make a cosmic time-line more accessible we can rescale everything so that the history of the Universe fits into a single Earth year. In so doing, each second that elapses in this illustrative cosmic calendar corresponds to a passage of about 445 years. Dates of astronomical significance in such a calendar would look something like the table below.

Clearly, much happened in a relatively short span of time as the Universe initially developed at a torrential pace. Things became somewhat more leisurely thereafter. From the human perspective, changes have once again become accelerated as man has begun to dominate the Earth in many respects. The entire recorded history of mankind represents a minuscule one part in a million of the age of the Universe. Interestingly, if we extend our calendar into the next "year," our Sun would survive in its present condition until late March, after which it would rapidly expand to become a red giant star, marking the demise of the Earth.

The cosmic year

event	date	time (hour:minute:second)
origin of the Universe	New Year's Day	00:00:00
era of inflation	New Year's Day	immediately
origin of the CBR	New Year's Day	00:26:27
galaxies begin to form	26 January	
Milky Way begins to form	21 February	
most distant known supernova explodes	5 June	
formation of the Earth	4 September	
Earth's oxygen atmosphere develops	9 November	
extinction of the dinosaurs	30 December	07:18 am
earliest written records to now	New Year's Eve	last 22 seconds
termination of the Sun	23 March of next year	

A leading idea, advocated by the American cosmologist Alan Guth in 1981, is that the Universe went through a period of inflation during which space briefly expanded at a much more rapid pace than now. This theory suggests that between the times of 10^{-34} and 10^{-32} seconds in the age of the Universe, space grew by a phenomenal factor of 10^{25}. Space would have expanded much more rapidly than even the speed of light. (That space might expand faster than the speed of light presents no physical contradictions; however, objects that are traveling in space cannot move faster than the speed of light.)

How does this theory help to explain anything? It means that before this period of inflation, everything in the Universe was tremendously much closer together, so that things that are widely separated now were not so in the past. This explains why the CBR looks the same at opposite sides of the Universe. The substantially closer proximity of these regions in the very early Universe allowed them to influence each other and become very similar in nature, thus producing the uniformity observed today.

Inflation also explains why space appears to be so flat. You could blow up a balloon to the size of your head, and it would look quite round. But suppose that you blew the balloon up to the proportions of the Earth, it would appear quite flat if you stood on it, and indeed that is how the ground of the Earth does look. So space may initially have been quite curved, but the incredible swelling effect of inflation has led to a Universe of flat space as seen today.

Inflation is thus a tidy concept for explaining certain features of the Universe not accounted for by the standard Big Bang model. And there are physical grounds for why inflation might occur. Theories for how forces such as electromagnetism and gravity came into being indicate that a period of inflationary expansion would follow as a result.

Inflation remains just a theory, however, and it cannot be observed. It occurred at a time when light was still bouncing off electrons, so it cannot be directly seen by looking to large distances when the Universe was young. Also, when inflation occurred, conditions in the early Universe were such that the temperature was around 10^{27} degrees, and such conditions are far beyond the capabilities of scientists to reproduce in experiments. Even so, inflationary theory does provide predictions about certain features of the Universe (such as the detailed splotchy brightness pattern of the CBR), and the results of recent experiments seem consistent with these predictions. But as astronomers continue to put inflation to the test, it may happen that an explanation just as good, or even better, may one day come along. That is the way with science.

Fate of the Universe

The Universe appears to be expanding, and there is a good deal of evidence in support of the Big Bang theory to account for this basic observation. So, in broad terms, astronomers believe that to some extent they understand where the Universe "has been." But what are the implications of expansion for the future of the Universe? The question may be addressed in two parts. What is the future of the expansion of the Universe? What is the future of the contents of the Universe?

In the case of the expansion, recall the simple analogy of a balloon in expansion, and how the size of the balloon in terms of its surface area increases as it expands. The Universe, however, is filled with mass, especially the elusive dark matter, and the gravity from this mass seeks to draw the Universe together. So there is a kind of competition between the space that seeks to grow and the matter that it carries, which tends to resist being separated.

The question that many cosmologists are seeking to address is, which one will win? The answer is sought by estimating all of the mass in the Universe as compared to the expansion of the Universe. A useful analogy is that of throwing a ball up in the air. The harder one throws, the higher it goes, and the longer it takes to fall back. There comes a point that if the ball is thrown hard enough, it never comes back. In this case the gravity of the Earth is always pulling on the ball, trying to make it fall down, but it is not sufficiently strong to make the ball stop moving away and to return. So it is with the Universe. If there is enough matter in the Universe, the gravity will be strong enough to stop the expansion, and a period of falling back or collapse will begin, ultimately to end in a Big Crunch. In our analogy, a Big Crunch would be like a ball that when thrown up, goes only so high, and then falls to smash into the ground. On the other hand, if there is not enough mass to cause a collapse, then the Universe will keep expanding forever.

At this time it appears that a forever expanding Universe may be the case. There is little to suggest that there is enough mass for gravity to halt the expansion and initiate contraction. Furthermore, the measurements of supernova explosions occurring at great distances in faraway galaxies indicate that the expansion of the Universe may actually be accelerating, meaning that the Universe is expanding faster now than it was in the past. This is rather different from what most astronomers had expected. It was thought that the Universe would expand forever, in the sense that when a ball is thrown just hard enough it never comes back, but only just so. (In this case the ball is always slowing down, but never falling back.) Instead, if the supernova results have been interpreted correctly, it may be that the Universe is more like a ball that has been thrown up with rocket blasters strapped around it to make it go faster and faster!

Supposing that the Universe just keeps expanding, the second issue is what happens to the constituents of the Universe, and when. Here we comment on the demise of stars, matter and black holes. The birth, life and death of stars is an ongoing cycle in galaxies that possess clouds of gas. The gas collapses under gravity to form a star. The star shines for a very long time, perhaps for billions of years, but eventually the hydrogen fuel that feeds its nuclear fusion "generator" runs out, and then the star dies. But during its life, the star will have given up some of its gas, perhaps through a wind or in a dramatic supernova explosion. However, the important fact is that the star does not give all of its mass back to be used in making new stars. A good fraction of the original gas that went into the star remains there in the form of a stellar corpse – a white dwarf, neutron star or black hole. This mass is trapped, never again to be used for making stars.

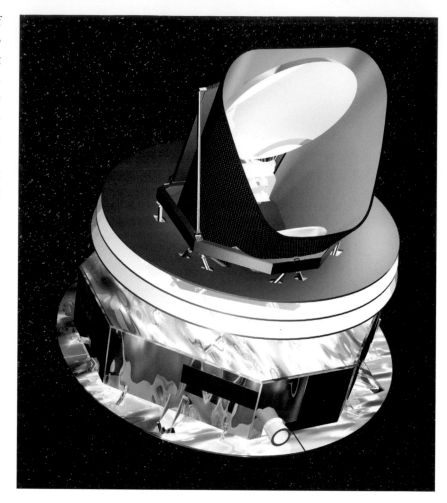

In a Universe like ours, where it is thought that no new gas is being spontaneously created anywhere, this cycle of making new stars cannot go on forever. Imagine lining up a great number of identical glasses, with only one glass being full of water. Taking this full glass, half the water is poured into the next glass. Of this, half is poured into the next, which now has just a quarter of the original. Continuing with this, there comes a point when there is no more water to pour into the next glass – the last drop has been reached. Similarly, there are diminishing returns in making new stars from the gas given up by old stars.

If the Universe is to continue existing for all time, then the formation of stars must stop at some point in the future. This is not to say that the stars will no longer exist, but rather that all of the stars will be the fading embers of once glorious luminaries. The blue and white stars will no longer be seen, and even the red stars will begin to blink out. Today, the Universe is thought to be about 14 billion years old. The last generation of stars will fade away when the Universe is around 100 trillion years old, or 10^{14} years, which is a very long time into the future.

Imagine that the Earth managed to survive for these many trillions of years, and that somehow time could be sped up so that we could watch this whole era. Many of the brightest stars in the night sky live for only a short time, perhaps less than a billion years. Some of them are already giants and will no longer be visible after 10 to 100 million years. It would not take long (only a few billion years) before familiar constellations dissolved, as most of the stars that can be seen today with the unaided eye will have "winked out." Of course, new stars will be made and will occasionally become part of new constellations. Eventually, however, there will be fewer and fewer new stars, the older ones will pass away and the gas necessary to make stars will get scarce. The long-lived, low-mass red stars shine so feebly that even though there are several quite close to Earth (less than 10 light-years away), most are not visible with the naked eye. These red stars would brighten as they evolved to become giants, but would rapidly vanish as they became tiny white dwarf stars, which become fainter with time. At this point the Universe will be a truly cold and dark domain.

Another issue concerning the constituents of the Universe is what will happen to the matter itself. Scientists are pursuing theories in an attempt to understand the nature of matter. Some of these ideas suggest that protons can decay or "break down" over extremely long periods of time, like 10^{32} years (100 million trillion trillion years). As the proton begins to break down into other kinds of bizarre particles and light, even the white dwarf and neutron star "corpses" will begin to go away, somewhat like the evaporation that occurs when a cup of water is left outside. However, the one type of stel-

lar corpse that is immune to this evaporation is the black hole, which brings us to the third epoch.

In chance encounters over great periods of time, the black holes will gradually sweep up any remaining planets, dead stars or other particles and gas debris in the Universe. But even the black holes will not last forever. The British physicist Stephen Hawking, at the University of Cambridge, has identified a mechanism by which black holes can evaporate over long periods of time, but for reasons unrelated to the decay of protons. The evaporation of black holes culminates in a violent explosion of intense gamma-ray light, much more powerful than X-rays. If the Sun were a black hole, it could survive for about 10^{65} years. Massive black holes with a million solar masses, like those thought to exist at the center of the Milky Way and many other galaxies, will live for about 10^{83} years. Black holes as massive as large galaxies (if any should ever become that big), could last for 10^{100} years.

At this point, perhaps we have pushed speculation far enough. Using only the physics currently known, the distant future of the Universe appears to hold a rather uninspiring eternity of loneliness and emptiness, not at all like the thriving cosmos that is seen around us today.

"Cosmo"-logue

Our journey has taken us from the familiar confines of near Earth orbit, where the Hubble Space Telescope resides, to the farthest reaches, in terms of both space and time, of the observable Universe. Astronomical discoveries in the last 100 years have propelled cosmology into a position at the forefront of scientific research, involving not only astronomers but also physicists and mathematicians from many different disciplines.

Significant advances include the discovery that the Universe is expanding, that it has a beginning in time, and that there exists a remnant cosmic background radiation from the raw and early Universe that is incredibly uniform across the sky. The Universe still appears to hold some surprises for us however – could the expansion of the Universe be speeding up, and if so what new physics might that reveal? There are clearly many more fascinating discoveries waiting just over the horizon.

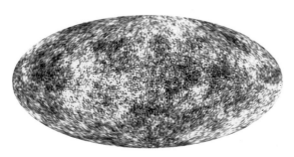

Life in the Universe

7

Our voyages in this book have led us to consider some very fundamental and important concepts, such as the origin of planets, the evolution of stars, the formation of galaxies, and the creation of the Universe itself. One fascinating and profound issue remains to be discussed, however, and that is whether life exists elsewhere in the Universe.

For thousands of years scientists and philosophers have gazed at the night sky and wondered whether there is life on planets around other stars. Today, the notion of extra-terrestrial life is fueled by our popular culture, through numerous science-fiction novels, films and television dramas. In this final chapter we will see that the quest for extra-terrestrial life is a subject of intense and serious scientific inquiry. Eminent, highly respected scientists throughout the world are currently involved in exciting experiments designed to probe the heavens for signs of life within and beyond the Solar System.

Earth is the only planet on which life is known to exist today, and there is no concrete proof so far that there is life elsewhere. However, our exploration of life in the Solar System and the Galaxy is very much in its infancy, and we certainly cannot eliminate the possibility of extra-terrestrial life. Indeed in the vastness of space, there is enough of a possibility for life to justify a dedicated search.

One of the humbling lessons covered in this book is that we do not hold a special location in the Universe. The Earth is not at the center of the Solar System, our Sun is located 26,000 light-years from the nucleus of our Galaxy, which in turn has no favored location in our local cluster of galaxies. This picture of mediocrity has been supplemented more recently by the scientific discovery of planets orbiting other stars, by evidence suggesting water in other locations of the Solar System, and by the (debated) possibility of fossil life within meteorites that have crashed on Earth from Mars.

It is very exciting to contemplate that we may be on the brink of discovering life elsewhere, and that within the next three decades we could perhaps witness one of the most important and profound discoveries of modern science. With this prospect in mind, we explore here how life started on Earth, and then discuss places where we might look for it in the Solar System. We also consider the chances of intelligent civilizations being present in our Galaxy and how we might look for them.

Before embarking further it should be noted that scientists do not always even agree on the definition of "life" itself. We limit ourselves in this chapter to life as we know it – life essentially based on carbon chemistry. Carbon is an excellent "building block"

▼ *The Pathfinder mission landed on Mars on 4 July 1997 and explored the surface of the planet for several weeks. This mission marked a new era of exploration of the Martian surface, with several experiments now scheduled over the next decade to search for primitive (bacterial) life forms.*

and is widespread in the Universe. In this definition, life changes with time and interacts with its environment. The key characteristics of living things on Earth are a capacity for metabolism, reaction to stimuli, growth and reproduction. These characteristics are in turn linked to a huge number of complex chemical reactions.

Life on Earth

The Earth today is extremely rich in life, with the number of different animal species alone coming to several million. We can use our planet to try to understand the conditions required for life to thrive and can then consider whether these conditions are entirely specific to Earth, or whether they are in fact general and thus applicable anywhere in the Universe.

It is remarkable to note how rapidly life emerged after the planet first formed about 4.5 billion years ago. Geological (fossil) records suggest evidence for the simplest forms of life more

▶ *This apparently grand view of the magnificent Hercules Cluster of galaxies in fact represents only a very tiny fraction of the Universe. There may be as many as a hundred billion galaxies in the observable Universe, each one typically made up of several billion stars. A vast number of these stars are much like our own Sun and it is likely that many stars are orbited by planets. There is, therefore, a very real possibility for the development of life elsewhere in such an extraordinarily large Universe.*

than 3.8 billion years ago. Based on these results some scientists argue that the earlier life began, the more inevitable it must have been. Although the precise events that led to the origin of life on Earth are not yet established, we do have some understanding of the most relevant processes.

At the earliest times after its formation, the Earth was mostly molten and was receiving intense radiation from the newly formed Sun. The planet cooled over millions of years, gradually acquiring a primitive atmosphere of methane, ammonia, hydrogen and water. Further additions were made by volcanic eruptions, which vented carbon dioxide, nitrogen and water vapor. Eventually, the Earth had cooled enough for liquid water to exist on its surface. Rain poured upon steaming rocks, and water accumulated into rivers and oceans. The warm seas, shallow lagoons and lakes were ideal for the development of organic compounds.

The young Earth's rich chemical environment was bombarded by powerful ultraviolet radiation from the Sun and by lightning storms that raged for millions of years. The energy from these sources caused a mixture of chemicals to rain into the oceans. Over a long time, more complex organic molecules formed, such as proteins and nucleic acids. In a process that is still not understood, these building blocks formed the very complex molecules known as ribonucleic acid (RNA) and deoxyribonucleic acid (DNA). These are the molecules that carry genetic codes to allow living creatures to reproduce.

Cells are the basic units of life, and collections of cells make up a living organism such as a human being. The initial development of microorganisms on Earth was very slow. Single-celled marine algae appeared as early as 3.5 billion years ago, but elaborate multicellular plant forms appeared only somewhat less than 1 billion years ago.

Many facets of the origin of life on Earth are still strongly debated, and some scientists have even suggested that life came to Earth from elsewhere in space, an idea known as panspermia. The basic notion of panspermia is that life did not originate on Earth but was seeded here in the form of bacteria and other organisms floating into our Solar System. The main argument against panspermia, though not necessarily definitive, is that the organisms could not survive the long journey between stars because the intense radiation in space would damage or destroy them. Besides, even if the concept of panspermia is correct, then the fundamental question of the origin of life is not answered, but simply relocated to another part of the Galaxy.

Most scientists agree that for life to exist and develop anywhere in the Universe, certain basic requirements must exist such as are found here on Earth, both now and in its past. One of these requirements is that some mechanism like RNA and DNA must operate for life to reproduce itself. Also, it would seem that any environment that is to be life-bearing should be situated near a star that acts as a source of radiant energy, and the environment must probably also possess liquid water, which is known to be crucial for so much of the Earth's biological activity. Although life on other planets, or even other moons, may be very different in appearance or operation from life on Earth, ample light and water and the capacity to procreate would seem to be essential.

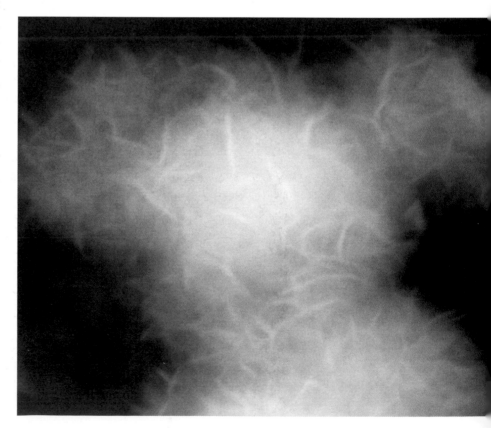

▼ *Life forms have now been discovered in some of the harshest environments on Earth. This greatly magnified image shows part of a colony of microbes found at depths of over 1000 m (3300 ft) in samples of ice collected in Antarctica. They have been buried there for several thousand years. Discoveries of fungi, algae and bacteria in deep-freeze conditions on Earth indicate that life is more robust than previously believed. This makes the possibility of life elsewhere, in our Solar System for example, more credible.*

Another important aspect of life on Earth, which may be pertinent in other places, is the role of the atmosphere. The air we breathe is about 70% nitrogen and 30% oxygen, but this has not always been so. Much of the hydrogen gas, which is what makes up most of the Sun, has long ago escaped from the Earth into space. Nitrogen gas has become more dominant through the decay of dead organisms and from volcanic activity. The substantial amounts of oxygen are thought to originate from plant life. Plants live and grow by means of photosynthesis, which requires carbon dioxide, water and sunlight. A by-product of the photosynthesis is that carbon dioxide is converted to oxygen, which is then released by plants into the atmosphere. The important thing is that life, in the form of plants, has led to a major adjustment of the Earth's environment, and this adjustment has been beneficial to humans since oxygen is necessary for us to live.

Another critical development in the evolution of our atmosphere that has permitted the existence of life on Earth is the lack of a major greenhouse effect. As discussed in Chapter 3, Venus suffers from a runaway greenhouse effect, so that its surface temperature can rise to around 500°C (930°F), which is more than hot enough to melt lead. This incredible heating is largely due to the thick carbon dioxide atmosphere of Venus, which acts to blanket in the heat of the planet. Mars also has a carbon dioxide atmosphere, but one that is much less dense, so it has only a rather mild greenhouse effect. It seems that the extensive oceans of the Earth have been critical in greatly reducing the amount of carbon dioxide in our atmosphere. Volcanoes release carbon dioxide gas in the same way as they release nitrogen; in fact they emit far more carbon dioxide than nitrogen. However, this carbon dioxide can dissolve in the oceans, eventually to be incorporated into carbon-based rocks. Thus the Earth maintains a relatively pleasant temperature that is suitable for life. The role played by the seas in this context is just another way that liquid water has proved to be important for life here.

In terms of intelligence, a critical stage in the history of human life on Earth has been the development of language, which not only allowed our ancestors to share ideas, but also enabled information to be stored as written documents to be passed down from generation to generation. Physically, humans have not changed substantially over the last tens of thousands of years; however, the capacity for intelligence has permitted, in a fairly short period of time, an almost complete dominance of humans over every other species of life on the planet. The past 100 years have seen immense leaps in technological advancements, with repercussions across the face of the globe. Our intelligence has had both positive and negative results, on ourselves, other species and the environment as a whole. Yet, it is this same intelligence that propels us to seek out life beyond our present confines.

The search for life in our Solar System

The Earth lies in a favorable "habitable" zone in the Solar System. It is at an ideal distance from the Sun for the surface of our planet to be at the right temperature for liquid water (oceans of it in fact), and for our climate to be relatively stable. The environment is perfect for long-term support of life as we know it. Scientists have discovered, however, that life is very robust and the previously assumed habitable zone for life is much broader. For example, biologists have discovered thriving microscopic organisms in the most extreme conditions on Earth, including the −15°C (5°F) deep-freeze of Antarctica, the boiling vents on the ocean floors, and at depths of more than 2 km (over a mile) in the darkest rocks. Many experts now consider that life could also have evolved, at least in its simplest forms, in other parts of our Solar System, despite the environments being very hostile.

Looking beyond our Earth, there are at least three locations in the Solar System that currently provide very exciting prospects of life forms sometime in their history. The sites of particular interest are Mars, a satellite of Jupiter called Europa, and the giant moon of Saturn named Titan. Each of these is a fantastic laboratory for the study of conditions that might have led to the formation and evolution of life, and we consider them in turn here.

Invaders from Mars?

The possibility that life once existed on Mars has fascinated writers for many years, from 19th-century claims of canals constructed by intelligent beings, to 20th-century tales of the invasion of Earth by octopus-like Martians! Scientific interest was significantly boosted in August 1996 by claims that a meteorite found in Antarctica, and thought to have originated on Mars, contained fossil remains and evidence for past microscopic life on the red planet. The finding remains controversial, and it will likely only be resolved by further studies of ancient terrain on the planet.

Mars is an exciting location since there is abundant evidence that liquid water has been present at the surface in the past and may even be present as a permafrost layer in the crust of the planet today. In 2001 NASA's Odyssey spacecraft detected hydrogen, indicating the presence of water ice in the upper meter (3 feet) of soil in a region surrounding Mars' south pole. At the present time Mars is too cold to have surface liquid water, and all its great canals were clearly made by rivers that flowed at least a billion years ago.

▼ *This image of the Cydonia region on the surface of Mars was taken by the Viking 1 spacecraft in 1976. It shows what appears to be a human face (above center), which led to spectacular claims that the "face" was constructed by Martian civilizations. Close-up observations taken by the Mars Global Surveyor in April 1998 confirmed the scientific view that the "face" and appearance of neighboring hills were due to a combination of naturally shaped plateaus, rock erosion and shadows.*

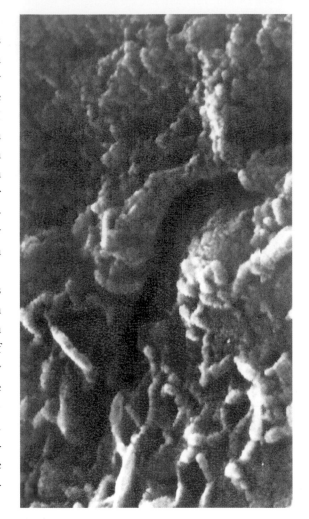

A series of spacecraft missions, scheduled over the next decade, aims to discover whether primitive life ever formed on Mars and whether it is still present today. The missions range from global reconnaissance and studies of water activity, to landers taking rock samples. They will include attempts to bring samples back to Earth for detailed analysis. One of the imminent exciting projects is the European Space Agency's Mars Express mission to be launched in 2003. It will carry, among other instruments, the British-led Beagle 2 Mars lander. Beagle 2 will seek evidence for past life by digging under the surface and chipping at rocks to test for the existence of organic matter.

Living oceans on Europa?

At first glance Europa may seem an unlikely place for life. The fourth-largest moon of Jupiter, it is slightly smaller than Earth's moon, and its

◀ *Tiny tube-like structures that may be possible fossils of bacteria-like organisms can be seen in this electron microscope image. They were found in an ancient meteorite, approximately 3.5 billion years old, that fell to Earth after being ejected from the surface of Mars as a result of an explosive impact. The largest possible fossils here are less than 1/100th of the width of a human hair.*

very bright surface is covered with ice, which is at a temperature of −160°C/−256°F. However, astronomers believe that the interior of Europa may be heated by a tidal tug-of-war between Jupiter's immense gravity and the gravity of Europa and Jupiter's other moons. Similar forces between the Earth and the Moon are strong enough to distort our oceans and produce the daily tides.

The heat generated in this manner could be keeping large parts of Europa's oceans slushy, in the upper layers, and perhaps liquid in the lower layers, which may be 100 km (60 miles) deep. This notion has been supported by some remarkably detailed photographs of Europa

taken by the Galileo satellite since December 1997. These photographs revealed evidence for features that looked like blocks of ice floating on slush; gaps could be seen where new crust seems to have formed between continent-sized plates of ice. Some areas on Europa contain fractured and rotated blocks of crust the size of small towns, with possible swirls of material between the fractured chunks.

The combination of liquid water, interior heat, and in-fall of organic matter from comets and meteorites means that Europa has the key ingredients for life. Whilst these ingredients do not, of course, guarantee life, it is worth remembering that life on Earth can be found in the

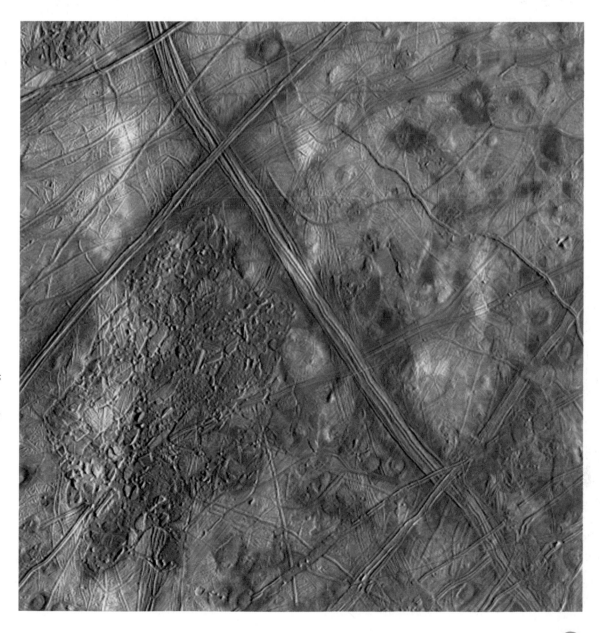

▶ *This detailed image of Europa's icy surface was taken from the Galileo spacecraft. The image covers an area of about 250 by 200 km (160 by 120 miles). It reveals a strange terrain of grooved ridges and crustal plates that appear to have broken apart and settled into new positions. The structures likely indicate subsurface water or slush, raising the remarkable possibility of organisms living on this eerie world.*

thick atmosphere. Of great interest is the fact that Titan's atmosphere today is the closest example we have to what Earth's environment may have been like billions of years ago when life began. Studies of Titan have shown that its atmosphere is mostly nitrogen, with thick organic (or carbon-based) chemicals, such as methane and ethane. These ingredients are similar to the primordial "soup" from which the first living organisms formed on Earth.

If all goes well with the complex Cassini mission, the Huygens probe will tell us much more about the chemicals in Titan's atmosphere as it descends at 6 km/s (almost 4 miles/s) toward its surface. At a height of 170 km (105 miles) above ground level, a huge parachute (9 m/30 feet across) will slow the descent, providing about $2\frac{1}{2}$ hours' worth of measurements before landing. It is hoped that the probe will survive the impact of landing for at least a few minutes so that we can learn if Titan's surface is gravel-like or icy. The probe may even land in lakes of liquid methane. Whatever Huygens finds, we stand to gain fantastic new insights into the history of our Earth.

It is clear that extensive chemical evolution occurred early in the history of the Solar System. We do not yet know, however, whether it progressed toward biological evolution anywhere else besides Earth. The next decade or so will witness the completion of several exciting spacecraft missions that should allow us to make a far more detailed pronouncement about life in the Solar System.

Looking for Earth-like planets

In Chapter 3 we saw how the search for other worlds beyond our Sun's family has been rejuvenated by the exciting discovery of giant (Jupiter-like) planets orbiting stars beyond our Solar System. The exciting challenge over the next two decades or so will be to find terrestrial extrasolar planets, which are perhaps 500 times less massive than Jupiter. To detect these Earth-sized planets scientists and engineers are planning to deploy in space an observation tool called an interferometer. Rather than using a single giant telescope, which would be very expensive to build and put in space,

▲ *This artist's depiction shows the European Space Agency's Huygens probe during its ambitious mission at Titan (scheduled for 2004). First the probe will enter the giant moon's dense upper atmosphere and use its heat shield to slow down. The shield will subsequently be jettisoned and parachutes used to complete the descent. Data will be collected during the whole journey and relayed to the orbiter above Titan.*

most bizarre and extreme sites. Europa is certainly an exciting environment, which awaits further detailed exploration.

Furthermore, magnetic readings of Jupiter's largest moon, Ganymede, taken by the Galileo spacecraft in December 2000, suggest there may also be a thick layer of melted salty water beneath Ganymede's icy crust.

Plunging into Titan

In October 1997 the National Aeronautics and Space Administration (NASA) and the European Space Agency (ESA) jointly launched an ambitious mission to explore the magnificent giant planet Saturn, together with its complex rings and some of its moons. The mission comprises an orbiting craft called Cassini (named after the astronomer who studied apparent "gaps" in Saturn's rings) and a probe called Huygens (named after the Dutch scientist who discovered Saturn's largest moon, Titan). Late in 2004, the Huygens probe is scheduled to make a remarkable descent to Titan's surface.

Titan is a mysterious world, larger than Mercury but smaller than Mars. The flyby of the Voyager 1 spacecraft in 1980, plus observations from Earth and the Hubble Space Telescope, have shown that Titan is completely shrouded by its

a space interferometer would consist of a few relatively small telescopes that are linked and operated in concert. The light collected by the separate telescopes is combined to produce a single image that is very sharp, and capable of resolving individual extrasolar planets. Over the next 15 to 20 years, new projects, such as the European Space Agency's Darwin mission, will help to discover Earth-like planets around other stars.

Of course, with billions of stars in our Galaxy, astronomers will have to decide which ones to observe most intensely. The star must not be too far away otherwise the planets might not be detected at all. The searches may also be restricted to middle-aged, yellow-white stars like our Sun. Stars that are much more massive than the Sun would not be suitable since these stars live fast and furious lives, with the result that there may not be enough time for life to develop. Earth-like planets will be rocky and close to their solar-like star, where the surface temperature is too high for the planet to be mostly made of ice. It must not be too close, however, since, like Venus, the planet would lose water from its atmosphere and surface.

A key probe for future space missions will be to examine the chemical make-up of the atmospheres, by taking spectrograms, in order to look for telltale signs of life. A terrestrial planet can only sustain an oxygen-rich atmosphere by having a continuous input of oxygen, since this element continually combines with other chemical elements on the planet's surface. The only known source to input large quantities of oxygen is oxygen-producing life, such as blue-green algae in the oceans. The oxygen can be detected as ozone produced in the upper-atmospheres of Earth-like planets. The exciting prospect raised by missions such as Darwin is that if we should detect life on several Earth-like planets, then it would be possible to directly study the progress of evolution.

Intelligence in the Universe

There are about 100 billion galaxies in the observable Universe, each containing tens or hundreds of billions of stars. Even if only a tiny fraction of these stars had planets, there would still be an unimaginably vast number of planets in the Universe. Among the most fascinating questions is whether some of these planets might support life that has evolved to develop an intelligent civilization, equal or more advanced than our own. Could there be aliens in space looking at our Galaxy or our star and discussing the Universe as we have done in this book? Nobody knows the answer. One concept that is sometimes applied in this context is the Copernican Principle, which states that we as observers in the Universe are neither special nor favored. Over the past several hundred years, any scientifically answerable questions regarding the uniqueness of our place in the Universe have returned the answer that we are not particularly special.

There is no solid proof at this moment, however, of extra-terrestrial intelligence, and it should be acknowledged that we could be the only advanced civilization around. Many astronomers find this an unacceptable conclusion though, citing the sheer numbers of stars and galaxies in a vast Universe as swaying the odds in favor of life elsewhere.

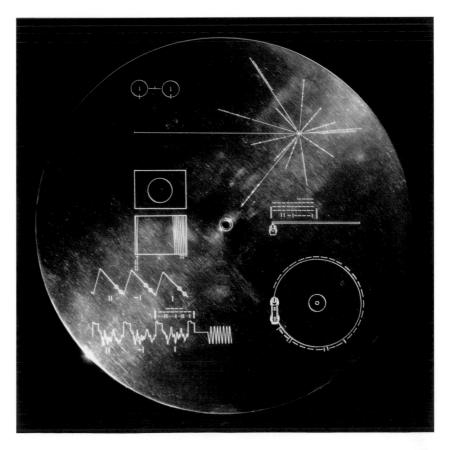

▼ Both of NASA's Voyager spacecrafts carried a gold-plated copper disk containing sounds and images selected to portray the diversity of life on Earth. The spacecrafts also carry small metal plaques which identify the place of origin of the crafts. As Voyager 1 and 2 escape entirely from our Solar System, the notion was that the plaques and records would announce our existence to other civilizations who might find them in the distant future.

In the absence of unambiguous and direct evidence, attempts may be made to estimate the number of extra-terrestrial civilizations in our Galaxy. In particular the American astronomer Frank Drake proposed, in 1961, a method of trying to account systematically for all the relevant unknown factors, in what amounts to an exercise of probabilities. Drake tried to provide a scientific basis for estimating the number of civilizations currently present in our Galaxy that are technologically capable of communicating with other Galactic civilizations. The number of communicating societies currently present is derived from the product of three astronomical factors, two biological factors and two sociological factors. As we shall see, a great deal of uncertainty exists in most of these terms.

The astronomical terms to consider in the Drake equation are the number of stars in our Galaxy (about 100 billion), the fraction of these that are similar to our Sun in size, temperature and lifetime (a modest estimate would be about 1 in 10), and finally the average number of planets per Sun-like star that are suitable for life (our Solar System would suggest about 1 in 10).

The exercise becomes more difficult as we consider the biological terms. The first factor is the fraction of suitable Earth-like planets on which life actually appears; it seems that at least the production of organic compounds was almost inevitable on Earth. The second biological factor is the fraction of these planets where life evolves to provide at least one intelligent species. This is a difficult term to tackle. It took over 4 billion years of evolution before modern man appeared on Earth. The fraction of these planets to evolve intelligent life is perhaps very high, as much as one maybe, since intelligence is of great benefit in evolution, and a clever species can perhaps survive better on the planet.

The final two factors in the Drake equation are sociological and even more difficult to guess. First, what is the fraction of intelligent societies in our Galaxy that would be technologically able and willing to attempt interstellar communications? For example, was humanity's rise to technological expertise inevitable? Alternatively, could an intelli-gent species remain as simple hunters and tool makers for their entire history? Nobody really knows the answer. If we wanted to be optimistic, we could adopt the view that all intelligent societies will eventually develop technologies, such as radio signals, capable of interstellar communication.

The final factor is the length of time any advanced civilizations continue to survive and communicate. Humans, for example, have been capable of sending and receiving radio signals for only a few decades out of a civilization period of more than 6000 years. How many more years might we survive? Our own destruction is sometimes feared as a result of nuclear wars, ecological and climatic disasters, and perhaps even impacts from comets. Maybe the most advanced civilizations somehow destroy themselves relatively soon after they develop, say within a few hundred years. In this case we would have very little hope of finding other civilizations in space. If, however, the average lifetime of a highly advanced species was instead very long, perhaps millions or even billions of years, then we would expect our Galaxy to be teeming with extra-terrestrial intelligence. In this case we should expect to succeed in making contact with them rather soon.

The Drake equation allows us to estimate the number of communicating civilizations in our Galaxy today. Interestingly, the number is strongly dependent on how long an advanced, radio-communicating civilization survives. So a search for extra-terrestrial civilization will only succeed if the lifetime of such a civilization is much longer than our own, by perhaps millions of years. It follows then that any contact we might make is likely to be with a civilization far more advanced than ours. This is both an exciting and a sobering thought as we consider next our current attempts to make communicative contact.

Listening out for E.T.

A possible way to search for other intelligent life forms is to attempt to communicate with them, and in particular to try to detect radio signals from them. The basic notion is that more advanced civilizations may have been transmitting signals for centuries and some of these may have reached our Solar System by now. Of course, although we regard radio transmissions as the basis for long-distance communication, highly advanced civilizations may use far more sophisticated energy sources for communicating signals, and we would be unaware of them and unable to detect them. Nevertheless, listening for radio signals from deep space is the usual method by which astronomers search for life. Their goal is to use radio telescopes to intercept and recognize messages from intelligent beings. This is a classic case of "searching for a needle in a haystack," since the Universe is awash with radio signals from astronomical sources, such as quasars, pulsars and planets, plus those emitted by terrestrial media broadcasts and air traffic control systems.

Radio astronomy developed from the use of radar employed in World War II, and it led, in 1960, to the first attempt to detect artificial signals from space. Led by Frank Drake, Project Ozma was carried out from West Virginia, United States, using a 26-m (85-feet) radio telescope. The instrument was pointed at two nearby stars, called Tau Ceti and Epsilon Eridani, which were similar to the Sun. After two months of listening, no signals were detected from alien beings, but nevertheless Project Ozma paved the way for more dedicated and sophisticated attempts. The general effort is broadly referred to as SETI (Search for Extra-terrestrial Intelligence).

During the 1960s, the Soviet Union dominated SETI with ambitious attempts to monitor large regions of the sky using antennae. The underlying philosophy was that at least a few highly advanced civilizations may be capable of radiating enormous transmitting power. In the early 1970s NASA began to participate, though its efforts were often dogged by critics, who regarded the SETI program as a waste of taxpayers' money. NASA's most ambitious SETI project used the

▼ *At 305 m (1000 ft) across, the Arecibo radio telescope is currently the largest single-dish telescope in the world. Built in a natural valley in the mountains of Puerto Rico, the Arecibo dish has occasionally been used to broadcast messages to regions of our Galaxy that might contain intelligent extra-terrestrial life. Any person on Earth can submit a proposal to request use of this giant telescope, though such requests are, of course, subject to competitive review!*

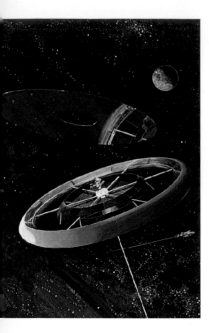

▲ *Could these be space colonies of the future? Shown here is an artist's fascinating portrayal of a huge space habitat where people live and work. When all the resources of our small planet have been exhausted, perhaps the human habitation of space will be a reality. The colony would, for example, draw energy from the Sun, and extract raw minerals plus water and oxygen from asteroids and the moons of Jupiter and Saturn. Having crowded the Solar System with space colonies, however, the human population may, within another few thousand years, have to travel interstellar distances to satisfy its increasing demand for essential resources. Could there be such a nomadic civilization already present in our Galaxy?*

305-m-wide (1000-feet) Arecibo radio telescope in Puerto Rico to listen for signals from the vicinity of the 1000 closest Sun-like stars. Within a year, however, the US Congress terminated funding since it was felt that the chances of success were too remote and uncertain.

With NASA no longer involved, the SETI Institute today is a privately funded endeavor, with its main activity being the US$3 million-a-year Project Phoenix. It retains the same basic strategy, which is to examine carefully the region around 1000 nearby solar-type stars. Project Phoenix is a considerably more comprehensive effort than any previously conducted. Various ongoing projects have detected signals that could fit the basic criteria for possible communication from extra-terrestrial civilizations. However, a given signal has never been detected twice, and until a signal is received repeatedly, it is not possible to claim that contact has been made. This is a very difficult task, demanding tremendous patience and effort, but the potential prize is truly spectacular.

If we were to succeed in establishing contact, a lot of difficult and interesting questions would be raised. Practical communication would be complex because of the enormous distances and times involved. Also, who would speak for planet Earth? Inviting attention from a considerably more advanced civilization than ours may hold the potential for good and bad for the human race. Should we be wary of answering back?

Has the Earth been visited?

If there are other civilizations in space, and if some of them are billions of years older and more advanced than us, then why is our Galaxy not colonized by them? Why has the Earth not been visited?

Some people believe that our planet has already been visited by extra-terrestrial beings. They point to "unidentified flying objects" (UFOs), for example, as proof. The "evidence" primarily consists of eyewitness accounts of those claiming to have seen them. Many sightings, even some backed by photographic materials, have later been admitted to be hoaxes. The majority of the remainder of sightings

are by honest people, though most often these cases can be attributed to natural or common sources, such as aeroplanes, balloons, meteors, flocks of birds, artificial satellites and so on. In fact, the planet Venus has prompted more reports of UFOs than any other source. Although there remain a few truly enigmatic sightings, very few scientists regard reports of UFOs as evidence for extra-terrestrial intelligence. Quite simply, extraordinary claims require extraordinary evidence. Many argue instead that any visits by aliens would be much less discrete and would leave no ambiguity.

If the Earth today is not being visited by extra-terrestrials, then there still remains the question as to whether interstellar explorers might have visited in the past. Perhaps their journeys are recorded in our historic records? This is a possibility, and some people cite as evidence pictures of "space gods" on ancient murals, pottery and rocks. The majority of scientists remain sceptical, however, arguing that detailed consideration of the human culture in question can most often provide a reasonable interpretation of the "mysterious" paintings, without the need to invoke extra-terrestrial visitors. No demonstrably authentic artifacts of alien visits are currently known to exist. Perhaps we have simply been overlooked by other forms of intelligent life in our Galaxy!

The profound question of whether mankind will come into contact with alien life forms remains unanswered. If advanced extra-terrestrial civilizations capable of long-distance space travel exist, then it is possible that they might one day visit us. Meanwhile, we must be cautious to suppose that our planet is the only one bearing life, and that the Earth commands a unique place in the Universe. We await that key message detected by a radio telescope, a sample of life from another part of our Solar System, or evidence for life-supporting atmospheres blanketing Earth-like extra-solar planets. It is impossible to predict what might be out there; we have to search in order to find out. All we know for certain is that we are here, and we are very small in a vast Universe that is rich in wonder and surprise. It is indeed a Universe in which so much still remains to be discovered and understood.

Glossary of terms

absolute zero
A temperature of –273°C (–460°F). Absolute zero is the coldest possible temperature.

accelerate
The way in which speed and direction of motion change with time. For example, an object that travels in a straight line with no variation in speed is not accelerating.

accretion
The gradual growth in mass, such as when a star forms by steadily accumulating gas.

accretion disk
A disk of matter spiraling in toward a massive object such as a star or black hole.

active galactic nucleus
The central nucleus of a galaxy that is unusually energetic and small, an example being a quasar.

angular diameter
The angle subtended by the diameter of an object as seen at a distance.

apparent brightness
The observed brightness of an astronomical object as measured by an observer on Earth.

asteroid
A small rocky body in space. Several thousand asteroids in the Solar System reside in the "belt" that lies between the orbits of Mars and Jupiter. Sometimes called a minor planet.

astrology
A system of belief that deals with the supposed effects on human destiny of the locations in the sky of the planets, Moon and Sun.

astronomical unit (AU)
The average distance between the Earth and Sun, equal to approximately 150 million km (93 million miles).

atom
Building block of matter. Each element (such as hydrogen or gold) is characterized by a unique type of atom.

aurora
A light display that results when charged particles from the solar wind enter the Earth's

atmosphere around the north or south poles. Sometimes called northern or southern lights. Aurorae have also been observed on other planets, such as Jupiter and Saturn.

axis
A straight line about which an object rotates or orbits.

Big Bang theory
A theory of cosmology to describe the expansion of the Universe, which is presumed to have begun with an explosive event.

binary star
Two stars that are gravitationally bound together and orbit one another. Many stars are found to be in binary-star systems.

black dwarf
The end state of the evolution of an isolated low-mass star that has exhausted all its energy sources and no longer emits detectable radiation.

black hole
An object or region of space with such strong gravitational pull that nothing (not even light) can escape it.

brown dwarf
A body that is more massive than a planet but not massive enough to undergo nuclear fusion of hydrogen in its core and to become a star. Brown dwarfs are "failed stars."

CBR
See cosmic background radiation

CCD
See charge-coupled device

celestial
Relating to the sky.

celestial pole
The projection of the north or south pole on to the celestial sphere.

celestial sphere
An imaginary globe surrounding the Earth, used to mark the apparent positions in the sky of astronomical objects, such as planets and stars.

center of mass
The average position of a

collection of massive bodies, weighted according to their distances from the center of mass.

Cepheid variable
A particular class of luminous pulsating stars with associated cyclic changes in brightness. The period of repetition is related to the true luminosity of the star, making these variable stars useful for measuring astronomical distances.

charge-coupled device (CCD)
An electronic device used to detect electromagnetic radiation at the focus of a telescope (or camera lens).

chromosphere
A part of the solar atmosphere, lying between the photosphere and the corona.

cluster of galaxies
A collection of galaxies that are gravitationally bound together. A cluster may contain several thousand galaxies.

comet
A small body made mostly of dust and ice in an elliptical orbit about the Sun. As a comet nears the Sun, some of its material boils off to form a long tail.

compound
A substance that can be further broken down into elements.

constellation
A grouping of stars identified on the celestial sphere. Many constellations are named after characters or beasts from ancient mythology.

core
The central part of an object, such as a planet, star or galaxy.

corona (of the Sun)
The tenuous and extremely hot outer atmosphere of the Sun.

cosmic background radiation (CBR)
A relic glow related to the Big Bang origin of the Universe. The CBR has a temperature of just 3 degrees above absolute zero and is very uniformly detected as microwave radiation coming from all directions in space.

cosmological constant
A term in the equations of general relativity that represents a repulsive (as opposed to attractive) force in the Universe.

cosmological principle
The assumption that at a given time observers who are located at different places in the Universe will see basically the same large-scale features.

cosmology
Study of the structure and evolution of the Universe as a whole.

crater
A bowl-shaped depression on the surface of a planet or moon, resulting from an impact of a smaller body.

crust
The outer surface layer of a terrestrial planet.

dark matter
An enigmatic component of the Universe. Dark matter has the attributes of mass and therefore gravity, but it gives off no light that can be detected at any part of the electromagnetic spectrum. Dark matter may constitute more than 90% of all mass in the Universe.

declination
An angular coordinate measure for the location of astronomical objects on the celestial sphere. It is similar to latitude for the Earth.

density
The amount of mass contained in a certain volume.

disk
A flattened rotating structure of gas and possibly dust.

Doppler effect
The change in color of light when the source of the light and the instrument of observation are moving with respect to one another. This is a similar effect to changes noted in the pitch of a whistle as a train passes by.

eclipse
An event during which one object passes in front of another.

eclipsing binary star
A binary star that is viewed from Earth such that one star is seen to pass in front and then behind the other during its orbit.

ecliptic
The apparent path of the Sun on the celestial sphere.

Edgeworth–Kuiper belt
A region lying beyond Neptune

and Pluto that is thought to contain large numbers of comets that orbit the Sun.

electromagnetic radiation
Another term for light. The radiation consists of waves moving through changing electric and magnetic fields.

electromagnetic spectrum
The complete range of light as characterized by wavelength, frequency or energy. Examples include X-ray light, ultraviolet light, visible light, infrared light and radio waves.

electron
A negatively charged subatomic particle that normally moves about the nucleus of an atom.

element
A substance that cannot be reduced by chemical processes into a simpler substance.

elliptical galaxy
A type of galaxy that appears oval shaped on the sky. Elliptical galaxies lack substantial amounts of interstellar gas and are mostly composed of older and redder stars.

equator
A circle on the surface of a sphere, 90° from each pole.

equinox
An intersection of the ecliptic and the celestial equator. Equinoxes mark the two times of the year when the length of the day and night are the same.

event horizon
The distance from which light can just escape from a black hole. The light will be trapped in the black hole if it is any closer than the event horizon.

extra-galactic
Anything beyond our Milky Way Galaxy.

flare
An explosive event occurring in or near an active region on the Sun.

force
Action on an object that causes its momentum to change.

frequency
Generally, the number of cycles occurring every second. For light, the number of electromagnetic waves passing a point every second.

fusion, nuclear
The process by which light atomic nuclei are combined into heavier ones, with a release of energy.

galactic cannibalism
A merger in which a larger galaxy consumes a smaller one.

galaxy
Collections of millions and billions of gravitationally bound stars, plus gas and dust.

Galaxy
The galaxy to which the Sun belongs. It is the same as the Milky Way Galaxy.

gamma rays
The most energetic form of electromagnetic radiation.

general relativity, theory of
A theory developed by Albert Einstein in the early 1900s to relate gravity, acceleration and the structure of space.

geocentric
With reference to the Earth being at the center.

giant (star)
An evolved phase of a star's life, when the core of a star has exhausted its supply of hydrogen for nuclear fusion. As a result, the star may swell to between 10 and 100 times the size of the Sun.

globular cluster
A collection of many thousands, and sometimes millions, of stars that are gravitationally bound and located in the halos of galaxies.

gravity
The mutual attraction between objects with mass. The greater the mass of a body, the stronger its gravitational pull.

gravitational lens
A chance alignment of objects in space such that one of the objects provides multiple images of the other by gravitationally bending its light.

greenhouse effect
The trapping by the atmosphere of some of the heat emitted by the surface of a planet, leading to an increase of the surface temperature.

halo (of galaxy)
The outermost regions of a galaxy, containing a roughly

spherical, sparse distribution of isolated stars and globular clusters.

heavy element
In astronomy, any element other than hydrogen or helium.

helio-
Prefix referring to the Sun.

heliocentric
With reference to the Sun being at the center.

homogeneous
A consistent and even distribution of matter that is the same everywhere.

Hubble constant
A factor that relates the distance of a galaxy to how fast it is receding from us. The value of the Hubble constant can be used to estimate the age of the Universe.

Hubble law
A law that relates the observed velocity of recession of a galaxy to its distance from us.

hypothesis
An idea or concept that is put forward, but remains to be tested with experiment or observation.

inflation (Universe)
A hypothetical phase in the very early Universe when space is thought to have expanded extremely rapidly for a short time.

infrared radiation
Part of the electromagnetic spectrum just beyond the red end of the visible range that can be seen by the eye.

interstellar medium
Gas and dust intermixed in the space between stars.

ionization
The process by which an atom loses one or more of its electrons, for example from bombardment by radiation.

irregular galaxy
A type of galaxy that has no obvious symmetry and does not adhere to the classes of spiral or elliptical galaxies.

isotropic
Appearing the same in every direction.

Jovian planet (giant planet)
Large planet resembling Jupiter,

mostly composed of gas (as opposed to rock and metal substances).

kinetic energy
Energy associated with motion.

Kuiper belt
See Edgeworth–Kuiper belt

laser
Device for amplifying a light signal at a particular wavelength into a coherent beam.

latitude
An angular coordinate measure for the location of objects on the Earth, either north or south of the equator.

light-year (ly)
The distance traveled by light in a vacuum in one year. One light-year is equal to 9460 billion km (5880 billion miles).

Local Group
The collection of galaxies that includes the Milky Way Galaxy and its nearest neighbors.

longitude
An angular coordinate measure for the location of objects on the Earth, measured either east or west of the Greenwich Meridian.

luminosity
The total energy emitted by a star each second at all wavelengths.

Magellanic Clouds
The Small and Large Magellanic Clouds are two irregular galaxies that orbit our own Galaxy. They can be seen from southern latitudes.

magnetic fields
Region of space associated with a magnetized object within which magnetic forces can be detected.

magnetic poles
Points on a magnetic body (such as the Earth) at which the density of magnetic lines is greatest.

magnitude
A logarithmic-based system for measuring the relative brightness of stars. A star with a smaller numerical value of magnitude is brighter than one with a larger value.

main-sequence
A reference to stars that are undergoing nuclear fusion of

hydrogen in their cores. A star spends most of its lifetime as a main-sequence star.

mantle (of Earth)
A layer of the Earth's interior that lies above the core and just below the crust.

mare (pl. maria)
A relatively dark colored and smooth region on the surface of the Moon that has few craters.

mass
A measure of the amount of matter contained within a body.

mass extinction
The catastrophic elimination of a species of life.

merger (of galaxies)
The result of two galaxies colliding to form a new galaxy.

meridian
A line on a sphere that connects the north and south poles and passes through a point directly above the observer.

Messier Catalogue
The catalog of nebulae, star clusters and galaxies compiled by Charles Messier in the late 18th century.

metals
In astronomy, an element other than hydrogen or helium. In the context of planets, the term "metals" has the more conventional meaning, referring to substances that are good conductors of electricity, such as iron, tin and so on.

meteor
A bright streak of light produced in the sky when a small piece of solid material from space enters the Earth's atmosphere and burns up. The event is often referred to as a shooting star.

meteorite
Any remains from a meteor that survives passage through the atmosphere and strikes the ground.

meteor shower
An event during which many meteors can be seen each hour, appearing to radiate from a common point in the sky.

microwave
Short-wave radio wavelengths.

Milky Way
The band of light that encircles the night sky. It is caused by the numerous stars and nebulae that lie close to the disk of our Galaxy.

minerals
The solid compounds that form rocks.

model
A theoretical construct used to explain an observation or experiment.

molecule
A particle that results from the combination of two or more atoms tightly bound together.

momentum
A measure of the state of motion of a body, defined as the product of its mass and velocity.

moon
See satellite

nebula
A cloud of gas and dust in space.

neutrino
A particle that travels at or near the speed of light, having no charge and little or no mass. Neutrinos are, for example, produced during fusion reactions in the cores of stars. They rarely interact with ordinary matter.

neutron
A subatomic particle with mass similar to a proton but no charge.

neutron star
A star of extremely high density made almost entirely of neutrons.

nova
Literally meaning "new star," a nova refers to a star (usually a white dwarf) that greatly brightens for a time as a result of an explosive event.

nuclear
Referring to the nucleus of an atom.

nucleus (of atom)
The heavy central part of an atom, consisting of protons and neutrons. It is orbited by one or more electrons.

nucleus (of galaxy)
The central region of a galaxy.

Oort cloud
A vast reservoir of comets at a distance of about a thousand billion kilometers from the Sun.

open cluster
A collection of typically hundreds or thousands of stars that are gravitationally bound together and located in the disk of the Galaxy. Open clusters are generally young and contain newly formed stars.

optical
In astronomy, term relating to the visible-light region of the electromagnetic spectrum.

orbit
The path of one astronomical body about another body or point.

ozone layer
A layer of gas in the Earth's atmosphere at a height of about 20–50 km (12–30 miles) above the surface, where incoming ultraviolet radiation from the Sun is absorbed.

photon
A discrete amount of light energy.

photosphere
The region of the Sun's atmosphere from which visible light escapes into space.

planet
A large body (rocky or gaseous) that orbits a star.

planetary nebula
The ejected outer layers of a red giant star, spread over a volume a few light-years across.

plasma
A gas that is fully or partially ionized.

positron
Similar to an electron but with opposite (positive) charge.

potential energy
Stored energy that can be converted into other forms of energy.

power
Energy output with time.

pressure
Amount of force that is spread over a given area.

prism
A glass object that splits white light into a spectrum.

prograde
Rotation or orbital motion in an anticlockwise direction when viewed from above. All the planets of the Solar System orbit in a prograde sense around the Sun.

prominence
A loop or sheet of glowing gas ejected from an active region above the Sun's photosphere.

proton
A subatomic particle with charge opposite to an electron and having a much greater mass.

protoplanet
A planet that is in the process of forming.

protostar
A collapsing mass of gas and dust out of which a star will be born.

pulsar
A spinning neutron star with strong magnetic fields that accelerate and eject high-energy particles. The resulting radiation is detected as short regular radio pulses at the Earth.

quasar
The nucleus of a very active galaxy. It appears star-like, but is at a great extra-galactic distance from us.

radar
The technique of bouncing radio waves off an object and then detecting the radiation that the object reflects back to the radio transmitter.

radiation
Usually refers to electromagnetic radiation, such as visible light, infrared, ultraviolet and so on.

radioactivity
The process by which certain kinds of atomic nuclei can spontaneously decay to become different nuclei, with the emission of subatomic particles and gamma rays.

radio telescope
A telescope designed to collect and detect radiation at radio wavelengths.

red giant
A large, cool star with a high luminosity, low surface temperature (around 3500°C/6300°F) and large proportions. The Sun will eventually become a red giant star.

reflecting telescope
A telescope that has a uniform concave mirror as its primary light gatherer.

refracting telescope
A telescope that uses glass lenses

to collect and focus light from astronomical sources.

retrograde
Rotation or orbital motion in a clockwise direction when viewed from above.

right ascension
An angular coordinate measure of astronomical objects on the celestial sphere. It is similar to longitude for the Earth.

satellite
Any smaller body orbiting a larger body. For example, the Moon is a satellite of the Earth.

SETI
Acronym for the Search for Extra-Terrestrial Intelligence. It is usually applied to organizations concerned with looking for signals from extra-terrestrial civilizations.

Seyfert galaxy
A spiral galaxy that has a small and unusually bright nucleus. It belongs to the class of galaxies that have active galactic nuclei (AGN).

solar eclipse
An eclipse of the Sun by the Moon, caused by the passage of the Moon in front of the Sun. Solar eclipses can only occur at the time of a new Moon.

Solar System
The bodies that are gravitationally bound to the Sun, such as the planets, moons, comets and asteroids.

solar wind
A stream of charged particles that escapes from the Sun's atmosphere at high speed and flows out into the Solar System.

solstice
The two times of the year when the Sun in the sky is farthest from the celestial equator.

spectral line
The radiation at a particular wavelength of light produced by the emission or absorption of energy by an atom.

spectroscopic binary star
Two stars revolving around a common center of mass that can be identified by cyclic changes in

the Doppler shift of the lines of their spectra.

spectrum
The array of colors or wavelengths apparent when light is dispersed, as by a prism.

speed
A measure of motion in terms of distance traveled over time.

spiral arms
The structures containing young stars and interstellar material that wind out from the central regions of some galaxies.

spiral galaxy
A type of galaxy in which most of the gas and stars are in a flattened disk that displays spiral arm structures. Our Milky Way Galaxy is a spiral galaxy.

star
A massive sphere of gas that shines by generating its own power.

star cluster
A collection of stars that are gravitationally bound to one another.

stellar evolution
The changes in structure and properties of a star over the course of time (usually over scales of millions to billions of years).

stellar wind
The outflow of gaseous material at high speed from a star.

Sun
The star about which the Earth is orbiting.

sunspot
A slightly cooler temporary region on the Sun's surface that appears dark by contrast to the surrounding hotter regions.

supercluster
A collection of several clusters of galaxies that spans a large region of space. The galaxies are not necessarily gravitationally bound to the supercluster.

supergiant
An evolved phase of stellar evolution. Supergiants are massive stars that can swell up

to several thousand times bigger than the Sun.

supernova
One of the brightest and most violent events in the Universe. It occurs when a star explodes. A Type Ia supernova occurs when a white dwarf accretes matter in a binary system. A Type II supernova results from the implosion of a massive star at the end of its life.

temperature
A measure of how hot or cold an object is; a measure of the average random speeds of microscopic particles in a substance. Temperatures may be quoted in degrees Celsius ($^\circ$C), degrees Fahrenheit ($^\circ$F) or Kelvin (K).

terrestrial planet
A planet that is predominantly composed of rocky and metal substances. Mercury, Earth, Venus and Mars are terrestrial planets.

theory
A set of laws and hypotheses that have been used to explain observed phenomena.

thermonuclear energy
The energy that results from encounters between particles that are given high velocities (through heating).

tidal force
The differences in the force of gravity across a body that is being attracted by another larger body. The result may be the deformation of the smaller body.

tides (on Earth)
The deformation of the ocean surface due to differences in the gravitational forces exerted by the Sun and the Moon.

ultraviolet radiation
Electromagnetic radiation of the region just outside the visible range, corresponding to wavelengths slightly shorter than blue light.

Universe
The total of all space, time, matter and energy.

variable star
A star with power output that changes over time.

velocity
The rate and direction in which distance is covered over some interval of time.

vernal equinox
The date on which the Sun crosses the celestial equator moving northward. It occurs on or near 21 March every year.

visible light
See optical

visual binary star
Two stars that revolve around a common center of mass and can both be seen through a telescope.

void
In astronomy, regions of space that appear to lack luminous matter. Superclusters of galaxies are separated by voids.

volume
A measure of the total three-dimensional space occupied by a body.

watt
A measure or unit of power output.

wavelength
The distance between two successive peaks or troughs of a wave.

weight
The total force on some mass due to gravitational attraction.

white dwarf
The collapsed end-state of a star that has exhausted its nuclear fuel supply and shines from residual heat.

X-rays
High-energy electromagnetic radiation, with photons of wavelengths intermediate between those of ultraviolet radiation and gamma rays.

year
The amount of time required for the Earth to complete one full orbit around the Sun.

zodiac
Traditionally the 12 constellations through which the Sun passes during the course of one year.

Index

Italic page numbers refer to captions

Acknowledgements

Endpapers NASA, H. Ford (JHU), G. Illingworth (USCS/LO), M.Clampin (STScI), G. Hartig (STScI), the ACS Science Team, and ESA; **Half–title** Philip's/Julian Baum; **Title** NASA, H. Ford (JHU), G. Illingworth (USCS/LO), M.Clampin (STScI), G. Hartig (STScI), the ACS Science Team, and ESA; **Foreword** ESO; **6–7** Courtesy SOHO/EIT consortium (SOHO is a project of International cooperation between ESA and NASA); **8** Simon Fraser/SPL; **9** NASA/JSC; **11** *top left* Courtesy SOHO/EIT consortium (SOHO is a project of International cooperation between ESA and NASA); **11** *top right* A. Dupree (CfA), STScI and NASA; **11** *bottom left* Hubble Heritage Team (AURA/STScI/NASA); **11** *bottom right* Rodger I. Thompson (University of Arizona), NASA/STScI; **13** Akira Fujii/DMI; **14** R. Scott; **16** *top left* Wally Pacholka; **16** *top right* Courtesy of HAO and Rhodes College. HAO is a division of the National Center for Atmospheric Research, which is supported by the National Science Foundation; **16** *center left* NASA/JPL/USGS; **16** *center right* Margarita Karovska (Harvard–Smithsonian Center for Astrophysics)/STScI; **16** *bottom left* STScI; **16** *bottom right* Hubble Heritage Team (AURA/STScI/NASA); **19** NASA/GSFC/DMSP; **20–21** NASA/Roger Ressmeyer/Corbis; **22** *top* ESO; **22** *bottom* ESO; **23** Philip's/Raymond Turvey; **24** NASA/KSC; **25** Corbis; **26** Bettmann/Corbis; **27** *left* NASA/JSC; **27** *right* STScI; **29** NASA/JSC; **30** NASA/JSC; **31** STScI; **32** *left* Corbis; **32** *center* Novosti; **32–33** Eugene A. Cernan/NASA/JSC; **33** *top* Corbis; **33** *center right* NASA/JPL; **34** Sun Courtesy SOHO/EIT consortium (SOHO is a project of International cooperation between ESA and NASA); **34** MERCURY NASA/JPL; **34** VENUS NASA/JPL; **34** EARTH NASA/JPL; **34** MARS NASA/JPL/USGS; **34** JUPITER NASA/JPL/University of Arizona; **34** SATURN NASA and the Hubble Heritage Team (AURA/STScI)/R.G. French (Wellesley College)/I. Cuzzi (NASA/Ames)/L. Dones (SwRI)/J. Lissauer (NASA/Ames); **34** URANUS NASA/STScI; **34** NEPTUNE NASA/JPL; **34** PLUTO Dr. R. Albrecht, ESA/ESO/ST-ECF/NASA; **36** Philip's/Julian Baum; **37** M. McCaughrean (MPIA)/C.R. O'Dell (Rice University)/NASA/STScI; **38** *left* NASA/JPL; **38** *right* NASA/JPL; **39** *top* NASA/JPL/UGS; **39** *bottom left* NASA/JPL; **39** *bottom right* NASA/JPL/USGS; **40** *top* Novosti; **40** *bottom* NASA/SPL; **41** *top* Bettmann/Corbis; **41** *bottom* NASA/JPL/USGS; **42** *top* Charles M. Duke Jr./NASA/JSC; **42** *bottom* NASA/JPL/USGS; **43** *top* NASA/JPL; **43** *bottom* NASA/JPL/Malin Space Science Systems; **44** *bottom* NASA/JPL/Cornell University; **45** *top* NASA/JPL/PIRL/University of Arizona; **45** *bottom* NASA/SPL; **46** NASA/JPL; **47** Corbis; **48** *top* STScI/NASA; **48** *bottom*

NASA/JPL; **49** *top* NASA/JPL; **49** *bottom* NASA/SPL; **50** Dr. R. Albrecht ESA/ESO/ST-ECF/NASA; **51** SwRI/Lowell Observatory/STScI/NASA; **52** NASA/JPL/USGS; **52** *right* NASA/JPL/USGS; **53** *top* Akira Fujii/DMI; **53** *bottom* European Space Agency/SPL; **54** NASA/STSci; **55** *top* Courtesy of HAO/SMM C/P Project Team and NASA. HAO is a division of the National Center for Atmospheric Research, which is supported by the National Science Foundation; **55** *bottom* Dr. Hal Weaver and T. Ed Smith (STScI) and NASA; **56** NASA/STSci; **57** NASA; **58** David Parker/SPL; **59** S.R. Heap (LASP/GSFC), NASA; **60** Philip's/Julian Baum; **62–63** J.P. Harrington and K.J. Borkowski (University of Maryland)/STScI/NASA; **63** Nik Szymanek (University of Hertfordshire); **64** *top* Institute of Space and Astronautical Science, Japan; **64** *bottom* Akira Fujii/DMI; **65** www.enviroweb.org; **66** *top* NASA/C.R. O'Dell and S.K. Wong (Rice University)/STScI; **66** *bottom* Jeff Hester and Paul Scowen (Arizona State University)/NASA; **67** NASA/ESA/Martino Romanello (ESO, Germany); **68** University of Bonn; **69** Philip's/Julian Baum; **70** Howard Bond (STScI)/Robert Ciardullo (Pennsylvania State University)/Bruce Balick (University of Washington)/Jason Alexander (University of Washington)/Arsen Hajan (US Naval Observatory)/Yervant Terzian (Cornell University)/Mario Perinotto (University of Florence)/Patrizio Patriarchi (Arcetri Observatory)/Vincent Icke (Leiden University)/Garalt Mellema (Stockholm University)/NASA ; **71** *top* Dan F. Figer (UCLA)/STScI/NASA; **71** *bottom* Jon Morse (University of Colorado)/STScI/NASA; **72** Hubble Heritage Team (AURA/STScI/NASA); **73** © Anglo-Australian Observatory. Photograph by David Malin; **74** *top left* Image courtesy Paul Scowen, Jeff Hester (ASU) and Mt. Palomar Observatories; **74** *top right* Jess Hester and Paul Scowen (Arizona State University)/NASA; **74** *bottom* NASA/CXC/SAO; **75** *top* Harvey Richter (University of British Columbia), STScI, NASA; **75** *bottom* Philip's/Raymond Turvey; **76** Philip's/Raymond Turvey; **78** Philip's/Julian Baum; **79** Corbis; **81** T. Nakajima and S. Kulikarni (CalTech)/S. Durrance and D. Golimowski (Johns Hopkins University)/STScI and NASA; **82** *top* A Caulet (ST-ECF, ESA)/NASA; **82** *bottom* STScI; **83** *top* John Trauger (JPL)/Jon Holtzman (New Mexico State University)/Hubble Heritage Team (AURA/STScI/NASA); **83** *bottom* STScI; **84–85** Jason Ware; **85** Hubble Space Telescope's WFPC Team; **86** STScI; **87** *top* Dr. Wendy L. Freedman, Observatories of the Carnegie Institution of Washington/NASA; **87** *bottom*

Peter Garnavich (Harvard-Smithsonian Center for Astrophysics)/the High-Z Supernova Search Team and NASA; **88** A. Dressler (Carnegie Institution of Washington)/M. Dickinson (STScI)/D. Macchetto (ESA/STScI)/M. Giavalisco (STScI)/NASA; **89** *top* Philip's/Raymond Turvey; **89** *bottom* COBE/DIRBE/NASA/GSFC; **90** *top* John Trauger (JPL)/Jon Holtzman (New Mexico State University)/Hubble Heritage Team (AURA/STScI/NASA); **90** *center* US Naval Research Laboratory; **91** NASA/U Mass./D. Wang, et al; **92** Hubble Heritage Team (AURA/STScI/NASA); **94** *top left* ESO; **94** *top right* ESO; **94** *center right* ESO; **94** *center left* © Anglo-Australian Observatory. Photograph by David Malin; **94** *bottom* R. Griffiths (Johns Hopkins University)/NASA/STScI; **96** *top* © Anglo-Australian Observatory/Royal Observatory, Edinburgh. Photograph by David Malin; **96** *bottom* © Anglo-Australian Observatory/Royal Observatory, Edinburgh. Photograph by David Malin; **97** Jason Ware; **98** *top* Brad Whitmore (STScI)/NASA; **98** *bottom* Kirk Borne (STScI)/NASA; **99** Hubble Heritage Team (AURA/STScI/NASA); **100** R. Williams (STScI)/NASA; **101** *top* Charles Steidel (CalTech), STScI, NASA; **101** *bottom* H. Ford and L. Ferrarese (Johns Hopkins University)/NASA; **103** John Bahcall (Institute for Advanced Study, Princeton)/ Mike Disney (University of Wales)/STScI; **104** *top* Philip's/Raymond Turvey; **104** *bottom* STScI; **105** W.N. Colley and E. Turner (Princeton University)/J.A. Tyson (Bell Labs, Lucent Technologies)/NASA/STScI; **106** Pieter van Dokkum/Marijn Pranx (University of Groningen)/ESA/NASA; **107** *left* Smithsonian Astrophysical Observatories; **107** *right* VIRGO Collaboration; **108–109** COBE/DIRBE/NASA; **111** The Starry Night, 1888 (oil on canvas) by Vincent van Gogh (1853–90) Musee d'Orsay, Paris, France/Bridgeman Art Library, Lauros/Giraudon/Bridgeman Art Library; **112** Philip's/Raymond Turvey; **113** Philip's/Raymond Turvey;**114** Jeffrey Newman (University of California at Berkeley)/STScI/NASA; **115** *top* K. Lanzetta and A. Yahil (SUNY)/NASA; **115** *bottom* M. Franx (Kapteyn Astronomical Institute)/G. Illingworth (Lick Observatory)/NASA; **116** *top* Jeff Hester (Arizona State University)/STScI/NASA; **116** *bottom* S. Maddox, W. Sutherland, G. Efstathiou, J. Loveday and G. Dalton (Astrophysics Department, University of Oxford); **118** *top* Roger Ressmeyer/Corbis; **118** *bottom* E.J. Ostrander/K.U. Ratnatunga/R. Griffiths (Carnegie Mellon University)/NASA; **119** COBE/NASA; **120** High-Z Supernova Search Team/HST/NASA; **121** Philip's/Raymond Turvey; **124** © Alcatel; **125** ESA; **126–127** NASA/JPL; **127** NOAO/AURA/NSF; **128** NASA/MSFC; **130** *top* NASA/GSFC; **130** *bottom* NASA/JPL; **131** NASA/JPL/University of Arizona; **132** ESA; **133** NASA/JPL; **135** David Parker/SPL; **136** NASA/JSC.